Mathematics 101

Practice Questions & Answers

by

PROFESSOR MATT MATIX

All rights reserved under the International Copyright Conventions

© 2017 MATT MATIX

Mathematics 101 Practice Questions and Answers

No parts of this book may be used or reproduced in any form, except for the inclusions of quotations embodied in critical articles and reviews, without the permission of the author and publisher.

Designed and printed in the United States of America.
ISBN - 13: 978-1974572267
ISBN - 10: 1974572269

1. EDITION

This book can be purchased for self-development and educational purposes

-- Copyrighted Material --

CONTENTS

Title	Page
Table of Contents	i

CHAPTER I

REVIEW OF ALGEBRA AND LINEAR EQUATIONS	1
Solving One-Variable Linear Equations	1
Formula Rewriting	5
Solving Radical Equations	7
Solving Exponential Equations	10
Solving Two-Variable Linear Equations	13
Binomial Expansions	19
Factoring	22
Quadratic Equation and Quadratic Formula	28
Practice Questions	31

CHAPTER II

PROBLEM SOLVING	43
Basic Short Problems	43
Formula Based Problems	47
Percentage Problems	50
Compound Interest	58
Practice Questions	60

CHAPTER III

CHART, TABLE and DATA ANALYSIS	67
Practice Questions	74

CHAPTER IV

FUNCTIONS	81
Practice Questions	93

CHAPTER V

INEQUALITIES, STATISTICS and COMPLEX NUMBERS	97
Inequalities	97
Statistical Analysis	102

Complex Numbers	105
Practice Questions	106

CHAPTER VI

COORDINATE GEOMETRY & BASIC SHAPES	109
Coordinate Geometry	109
Slope of a Line	110
Equation of Lines: Knowing one point and slope	110
Equation of Lines: Knowing two points	111
Parabolas	121
Shape Geometry	124
Practice Questions	125
Key to Practice Questions	131

CHAPTER I

Review of Algebra and Linear Equations

Example 1.1   Solving One-Variable Linear Equations

Find the value of x; if 3x − 5 = x + 9

Solution:

3x − 5 = x + 9	3x − 5 = x + 9
3x − x = 9 + 5 or	−5 − 9 = x − 3x
2x = 14	−14 = −2x
x = 7	7 = x

Example 1.2

Solve $\frac{5x}{4}$ + 20 = 3x + 13

Solution:

$\frac{5x}{4}$ − 3x = 13 − 20	20 − 13 = 3x − $\frac{5x}{4}$
$\frac{5x-12x}{4}$ = −7 or	7 = $\frac{12x-5x}{4}$
$\frac{-7x}{4}$ = −7	7 = $\frac{7x}{4}$
−7x = −28	28 = 7x
x = 4	4 = x

Example 1.3

Solve $\dfrac{5x+1}{4} - \dfrac{4x+6}{3} = 3x - 7$

Solution:

$\dfrac{5x+1}{4} - \dfrac{4x-6}{3} = 3x - 7$ $\qquad\qquad$ $\dfrac{5x+1}{4} - \dfrac{4x-6}{3} = 3x - 7$

$\quad(3)\quad\ \ (4)\quad\ (12)$ $\qquad\qquad\qquad\quad(3)\quad\ \ (4)\quad\ (12)$

First, we need to clear the denominators by equating them to the Least Common Denominator which is 12. So, we will expand each denominator to attain 12. Be aware that $(3x - 7)$ has a denominator of 1.

$\dfrac{15x+3}{12} - \dfrac{16x-24}{12} = \dfrac{36x-84}{12}$ $\quad$ or $\quad$ $\dfrac{15x+3}{12} - \dfrac{16x-24}{12} = \dfrac{36x-84}{12}$

15x + 3 − 16x + 24 = 36x − 84 $\qquad\qquad$ 15x + 3 − 16x + 24 = 36x − 84

−x + 27 = 36x − 84 $\qquad\qquad\qquad\qquad$ 27 − x = 36x − 84

111 = 37x $\qquad\qquad\qquad\qquad\qquad\qquad$ −37x = −111

3 = x $\qquad\qquad\qquad\qquad\qquad\qquad\qquad$ x = 3

Example 1.4

Solve $\dfrac{3m-8}{5} - 1 = \dfrac{2m-9}{3}$

Solution:

$\dfrac{3m-8}{5} - 1 = \dfrac{2m-9}{3}$ $\qquad\qquad$ $\dfrac{3m-8}{5} - \dfrac{2m-9}{3} = 1$

$\quad(3)\quad (15)\quad (5)$ $\qquad$ or $\qquad\quad(3)\quad\quad (5)$

$\dfrac{9m-24}{15} - \dfrac{15}{15} = \dfrac{10m-45}{15}$ $\qquad\qquad$ $\dfrac{9m-24}{15} - \dfrac{10m-45}{15} = 1$

9m − 24 − 15 = 10m − 45 $\qquad\qquad$ $\dfrac{9m-24-10m+45}{15} = 1$

−39 + 45 = m $\qquad\qquad\qquad\qquad\quad$ 21 − m = 15

m = 6 $\qquad\qquad\qquad\qquad\qquad\qquad\quad$ 6 = m

Review of Algebra and Linear Equations

Example 1.5

Solve $\quad \left(\frac{6}{5} - \frac{3}{4}\right)\left(5 + \frac{5}{3}\right) = x$

Solution: $\quad \left(\frac{6}{5} - \frac{3}{4}\right)\left(\frac{5}{1} + \frac{5}{3}\right) = x$

$\qquad\qquad$ (4) $\;$ (5) $\;$ (3)

$\qquad \left(\frac{24 - 15}{20}\right)\left(\frac{15+5}{3}\right) = x$

$\qquad \left(\frac{9}{20}\right)\left(\frac{20}{3}\right) = x$

$\qquad \frac{9}{3} = x$

$\qquad x = 3$

Example 1.6

Solve $\quad \dfrac{\frac{3}{4} + \frac{2}{3}}{\frac{4}{3} + \frac{3}{2}} = \dfrac{1}{x}$

Solution: $\quad \dfrac{\frac{9+8}{12}}{\frac{8+9}{6}} = \dfrac{1}{x}$

$\qquad \dfrac{\frac{17}{12}}{\frac{17}{6}} = \dfrac{1}{x}$

$\qquad \dfrac{17}{12} \cdot \dfrac{6}{17} = \dfrac{1}{x}$

$\qquad \dfrac{1}{2} = \dfrac{1}{x}$

$\qquad x = 2$

> **Remember!**
>
> Division is the inverse of multiplication.
>
> $\dfrac{\frac{17}{12}}{\frac{17}{6}} = \dfrac{17}{12} \cdot \dfrac{6}{17}$
>
> Just flip the denominator and multiply it with the numerator

Example 1.7

Solve $\quad \dfrac{4}{9-\frac{49}{9}} - \dfrac{1}{8} = x$

Solution $\quad \dfrac{4}{\frac{81-49}{9}} - \dfrac{1}{8} = x$

$\dfrac{4}{\frac{32}{9}} - \dfrac{1}{8} = x$

$4 \cdot \dfrac{9}{32} - \dfrac{1}{8} = x$

$\dfrac{9}{8} - \dfrac{1}{8} = x$

$\dfrac{8}{8} = 1 = x$

Example 1.8

Solve $\quad 2t - (3t-4) + t/2 + 10 = t - (5t+1)/3 + 14$

Solution: $\quad 2t - 3t + 4 + \dfrac{t}{2} + 10 = t - \dfrac{5t+1}{3} + 14$

$-t + \dfrac{t}{2} + 14 = \dfrac{3t-5t-1}{3} + 14$

$\dfrac{-2t+t}{2} = \dfrac{-2t-1}{3}$

$\dfrac{-t}{2} = \dfrac{-2t-1}{3}$

We can do cross multiplication here.

$-3t = -4t - 2$

$t = -2$

Review of Algebra and Linear Equations

Example 1.9

Solve $k - (2k + 3)2 + k/3 + k/2 = 2k - (5k + 1)/3 - (k - 1 - 3k)3 - 5$

Solution: $k - 4k - 6 + \dfrac{k}{3}_{(2)} + \dfrac{k}{2}_{(3)} = 2k - \dfrac{5k+1}{3} - 3k + 3 + 9k - 5$

$$-3k - 6 + \dfrac{2k+3k}{6} = 8k - \dfrac{5k+1}{3} - 2$$

$$\dfrac{5k}{6} = 11k + 4 - \dfrac{5k+1}{3}$$

$$\phantom{\dfrac{5k}{6}}\;_{(3)}\;\;\;_{(3)}$$

$$\dfrac{5k}{6} = \dfrac{33k + 12 - 5k - 1}{3}$$

$$\dfrac{5k}{6} = \dfrac{28k + 11}{3}$$

> **Be Careful !!** Multiplication/ Division comes before Addition/Subtraction.

Simplify further by dividing both denominators into 3.

$$\dfrac{5k}{2} = \dfrac{28k + 11}{1}$$

Do the cross multiplication.

$5k = 56k + 22$

$-51k = 22$

$k = -\dfrac{22}{51}$

Example 1.10   Formula Rewriting

The formula below shows the relationship between A, B and C.

$$A = \dfrac{2B + C + 4}{3}$$

How would you rewrite the formula that gives B in terms of A and C?

Solution:

$$A = \frac{2B + C + 4}{3}$$

$$3A = 2B + C + 4$$

$$3A - C - 4 = 2B$$

$$\frac{3A - C - 4}{2} = B$$

$$B = \frac{3A - C - 4}{2}$$

Example 1.11

The formula below is frequently used by investment bankers to assess W_0, the present value of initial endowment, where y_0 is the sum of current income, y_1 is the present value of income at the end of the period and r is the market-determined interest rate.

$$W_0 = y_0 + y_1(1+r)^{-1}$$

How would you rewrite the formula that gives the y_1 in terms of r, W_0 and y_0?

Solution:
$$W_0 = y_0 + y_1(1+r)^{-1}$$

$$W_0 = y_0 + \frac{y_1}{1+r}$$

$$W_0 - y_0 = \frac{y_1}{1+r}$$

$$(1+r)(W_0 - y_0) = y_1$$

$$\mathbf{y_1 = (1+r)(W_0 - y_0)}$$

Review of Algebra and Linear Equations

Example 1.12 Solving Radical Equations

Solve $x = 2\sqrt{3}(\sqrt{12} + 3\sqrt{3} - 2\sqrt{75})$

$x = 2\sqrt{3}(\sqrt{3.4} + 3\sqrt{3} - 2\sqrt{3.25})$

> **Warning!!** We are using period (.) as a sign of multiplication.

$x = 2\sqrt{3}(2\sqrt{3} + 3\sqrt{3} - 10\sqrt{3})$

> **Watch out!!** Inside the Parentheses must be handled before anything else

$x = 2\sqrt{3}(-5\sqrt{3})$

$x = -10.3$

$x = -30$

Example 1.13

Solve $x + 2 = \sqrt{x^2 + 2x + 10}$

Solution: $(x + 2)^2 = x^2 + 2x + 10$

$x^2 + 4x + 4 = x^2 + 2x + 10$

$2x = 6$

x = 3

Example 1.14

Solve $\dfrac{3x}{5} = \dfrac{3}{\sqrt{5}}$

Solution $\dfrac{3x}{5} = \dfrac{3}{\sqrt{5}}$

$(\sqrt{5})$ .... Expand both numerator and denominator by $\sqrt{5}$

$\dfrac{3x}{5} = \dfrac{3\sqrt{5}}{5}$

$$3x = 3\sqrt{5}$$

$$x = \sqrt{5}$$

Example 1.15

Solve $\quad \dfrac{\sqrt{48}}{\frac{1}{\sqrt{3}}+\frac{1}{\sqrt{27}}} = x$

Solution $\quad \dfrac{\sqrt{16 \cdot 3}}{\frac{1}{\sqrt{3}}+\frac{1}{\sqrt{3 \cdot 9}}} = x$

$$\dfrac{4\sqrt{3}}{\frac{1}{\sqrt{3}}+\frac{1}{3\sqrt{3}}} = x$$

(3)

$$\dfrac{4\sqrt{3}}{\frac{3}{3\sqrt{3}}+\frac{1}{3\sqrt{3}}} = x$$

$$\dfrac{4\sqrt{3}}{\frac{4}{3\sqrt{3}}} = x$$

$$4\sqrt{3} \cdot \dfrac{3\sqrt{3}}{4} = x \qquad \boxed{\textbf{Be Practical!} \text{ The 4s are cancelling each other out and } (\sqrt{3}) \cdot (\sqrt{3}) = 3}$$

$$9 = x$$

Example 1.16

Solve $\quad \dfrac{\sqrt{12}}{\sqrt{27}+\frac{1}{\sqrt{3}}} = x^{-1}$

Solution: $\dfrac{\sqrt{3}\cdot\sqrt{4}}{3\sqrt{3}+\dfrac{1}{\sqrt{3}}} = x^{-1}$ $\boxed{\sqrt{27} = \sqrt{3\cdot 3^2} = 3\sqrt{3}}$

$\dfrac{2\sqrt{3}}{3\sqrt{3}+\dfrac{1}{\sqrt{3}}} = x^{-1}$

$(\sqrt{3})$

$\dfrac{2\sqrt{3}}{\dfrac{9}{\sqrt{3}}+\dfrac{1}{\sqrt{3}}} = x^{-1}$

$\dfrac{2\sqrt{3}}{\dfrac{10}{\sqrt{3}}} = \dfrac{1}{x}$

$2\sqrt{3}\cdot\dfrac{\sqrt{3}}{10} = \dfrac{1}{x}$

$\dfrac{3}{5} = \dfrac{1}{x}$

$x = \dfrac{5}{3}$

Example 1.17

Solve $\dfrac{1}{2-\sqrt{5}} - \dfrac{1}{2+\sqrt{5}} = x\sqrt{5}$

Solution: $\dfrac{1}{2-\sqrt{5}} - \dfrac{1}{2+\sqrt{5}} = x\sqrt{5}$

$(2+\sqrt{5}) \quad (2-\sqrt{5})$
.... $(2+\sqrt{5})(2-\sqrt{5}) = 4 - 2\sqrt{5} + 2\sqrt{5} - 5$
$\dfrac{2+\sqrt{5} - (2-\sqrt{5})}{4-5} = x\sqrt{5}$

$\dfrac{2+\sqrt{5} - 2+\sqrt{5}}{-1} = x\sqrt{5}$

$-2\sqrt{5} = x\sqrt{5}$

$x = -2$

Example 1.18   Solving Exponential Equations

Solve $3^3 \cdot 3^5 \cdot 3^4 = 9^x$

We need to apply some of the basic laws of exponents to our equation,

Such as: $9^x = 3^{2^x} = 3^{2x}$

So, $3^3 \cdot 3^5 \cdot 3^4 = 3^{2x}$

$$3^{12} = 3^{2x}$$

$$2x = 12$$

$$x = 6$$

Example 1.19

Solve $\dfrac{27x^9}{x^2} = 9^5$

Solution:

$$3^3 \cdot x^{9-2} = 3^{10}$$

$$x^7 = 3^{10-3}$$

$$x^7 = 3^7$$

$$x = 3$$

Example 1.20

Solve $\dfrac{2^3 \cdot 5^2 + 2^2 \cdot 5^3}{35} = 4x$

Review of Algebra and Linear Equations

Solution:

$$\frac{2^2 \cdot 5^2(2+5)}{35} = 4x$$

$$\frac{2^2 \cdot 5^2}{5} = 4x$$

$$2^2 \cdot 5 = 4x$$

$$x = 5$$

Example 1.21

Solve $\quad \dfrac{9^4 \cdot 6^{-8}}{4^{-6}} = x$

Solution: $\quad \dfrac{9^4 \cdot 4^6}{6^8} = x$

$$\frac{3^{2 \cdot 4} \cdot 2^{2 \cdot 6}}{2^8 \cdot 3^8} = x$$

$$\frac{3^8 \cdot 2^{12}}{2^8 \cdot 3^8} = x$$

$$2^{12-8} = x$$

$$2^4 = x$$

$$x = 16$$

> If you have a minus sign "−" in an exponent, you can throw the entire number under the division line by changing the sign of the exponent.
>
> For ex: $8^{-3} = \dfrac{1}{8^3}$; $\dfrac{2^{-5}}{3^{-7}} = \dfrac{3^7}{2^5}$

Example 1.22

What is the simplified form of the equation below?

$$\frac{a^4 + a^2 b - a^2 b^2 - b^3}{a^3 + ab - a^2 b - b^2} =$$

Solution:

$$\frac{a^2(a^2+b) - b^2(a^2+b)}{a(a^2+b) - b(a^2+b)} =$$

$$\frac{(a^2-b^2)(a^2+b)}{(a-b)(a^2+b)} =$$

$$\frac{(a-b)(a+b)}{(a-b)} = a + b$$

Example 1.23

$$\frac{x^{1/3} \cdot y^{-2}}{x^{-2/3} \cdot y^{-1}} = ?$$

Solution:

We need to summon all *x* exponents above *x*; and all *y* exponents above *y*.

Let's do the exponents of *x* first:

$$\frac{x^{\frac{1}{3}+\frac{2}{3}} \cdot y^{-2}}{y^{-1}} = \frac{x^{\frac{3}{3}} \cdot y^{-2}}{y^{-1}} =$$

Now, let's do the exponents of *y* (*throw y^{-2} under the division line*):

$$\frac{x^1}{y^{-1+2}} = \frac{x}{y}$$

Example 1.24

$$\frac{x^{\frac{5}{3}} \cdot y^{\frac{-2}{5}}}{x^{-1} \cdot y^{\frac{2}{3}}} = ?$$

Review of Algebra and Linear Equations

Solution:

$$\frac{x^{\frac{5}{3}+1} \cdot y^{\frac{-2}{5}}}{y^{\frac{2}{3}}} = \frac{x^{\frac{8}{3}} \cdot y^{\frac{-2}{5}}}{y^{\frac{2}{3}}} =$$

$$\frac{x^{\frac{8}{3}}}{y^{\frac{2}{3}+\frac{2}{5}}} = \frac{x^{\frac{8}{3}}}{y^{\frac{10+6}{15}}} = \frac{x^{\frac{8}{3}}}{y^{\frac{16}{15}}} =$$

Remember that:

$$x^{\frac{m}{n}} = \sqrt[n]{x^m}$$

$$\frac{\sqrt[3]{x^8}}{\sqrt[15]{y^{16}}} = \frac{\sqrt[3]{x^6 \cdot x^2}}{\sqrt[15]{y^{15} \cdot y^1}} =$$

$$\frac{x^{\frac{6}{3}}\left(\sqrt[3]{x^2}\right)}{y^{\frac{15}{15}}\left(\sqrt[15]{y^1}\right)} = \frac{x^2 \sqrt[3]{x^2}}{y^{15}\sqrt[15]{y}}$$

Example 1.25   Solving Two-Variable Linear Equations

If $\frac{3}{2} = \frac{4x}{3y}$, what is the value of y in terms of x?

Solution:

$$\frac{3}{2} = \frac{4x}{3y}$$

Let's do the cross multiplication here.

$$9y = 8x$$

And leave y alone.

$$y = \frac{8x}{9}$$

Example 1.26

If $\dfrac{3k}{2} = 2h$, what is the value of $\dfrac{h}{k}$?

Solution:

$$\dfrac{3k}{2} = 2h$$

$$3k = 4h$$

In order to find $\dfrac{h}{k}$, we can divide both sides by 4k,

or

We can first divide both sides by 4, and then divide them by k.

$$\dfrac{3k}{4} = \dfrac{4h}{4}$$

$$\dfrac{3k}{4} = h$$

$$\dfrac{3k}{4k} = \dfrac{h}{k}$$

$$\dfrac{h}{k} = \dfrac{3}{4}$$

Example 1.27

If $\dfrac{2x+5}{3y+1} = 4$, what is the value of y in terms of x?

Solution:

$$\dfrac{2x+5}{3y+1} = 4$$

$$2x + 5 = 12y + 4$$

Review of Algebra and Linear Equations

$$2x + 1 = 12y$$

$$y = \frac{2x+1}{12}$$

Example 1.28

If (a, b) satisfies the system of equations below, what is the value of a?

$$a - 2b = 7$$
$$b - a = -4$$

Solution: We need to align the variables (a goes under a, and b goes under b).

$$a - 2b = 7$$
$$-a + b = -4$$
$$+ \text{---------------------}$$

Now, we can add the equations up.

$$a - a - 2b + b = 3$$

$$-b = 3$$

$$b = -3$$

Finally, we need to plug b (-3), into any one of the equations to find a.

$a - 2(-3) = 7$	$b - a = -4$
$a + 6 = 7$	$-3 - a = -4$
$a = 1$	$a = 1$

or

Example 1.29

If (x, y) satisfies the system of equations below, what is the value of $y - x$?

$$2x + 3y = 31$$
$$y + x = 12$$

Solution: Let's align them and expand one of the equations in a way to get rid of one of the variables.

$$2x + 3y = 31$$
$$-2\,(x + y = 12)\,-2$$

$$2x + 3y = 31$$
$$-2x - 2y = -24$$
+-----------------------
y = 7

x + 7 = 12

x = 5

y − x = 7 − 5 = **2**

Example 1.30

If (x, y, z) satisfies the system of equations below, what is the ratio of x to y?

$$\frac{x+y}{z} = \frac{4}{3}$$

$$\frac{y}{z} = \frac{4}{5}$$

$$\frac{x}{y} = ?$$

Solution:

$$4z = 3x + 3y$$

$$4z = 5y$$

$$5y = 3x + 3y$$

$$2y = 3x \qquad \leftrightarrow \qquad \frac{x}{y} = \frac{2}{3}$$

Review of Algebra and Linear Equations

Example 1.31

If (x, y, z) satisfies the system of equations below, and of $x > 0$, $y > 0$, and $z > 0$, what is the value of $(x + y + z)$?

$$\frac{y+z}{x} = \frac{5}{4}$$

$$xz + xy = 45$$

Solution:

$$\frac{y+z}{x} = \frac{5}{4}$$

$$x(z + y) = 45$$

$$y + z = \frac{45}{x}$$

$$\frac{\frac{45}{x}}{x} = \frac{5}{4}$$

$$\frac{45}{x^2} = \frac{5}{4}$$

| Let's simplify! |

$$\frac{9}{x^2} = \frac{1}{4}$$

$$x^2 = 36$$

$$x = \pm 6$$

| Since $x > 0$, $x = 6$ |

$$y + z = \frac{45}{x} = \frac{45}{6} = \frac{15}{2}$$

$$x + y + z = 6 + \frac{15}{2}$$

$$x + y + z = \frac{27}{2}$$

Example 1.32

If (a, b) satisfies the system of equation below, what is the value of $a + b$?

$$16^a \cdot 9^a = 6^b \cdot 8^2$$

Solution:

$$2^{4a} \cdot 3^{2a} = 2^b \cdot 3^b \cdot 2^6$$

$$2^{4a-b-6} = 3^{b-2a}$$

Both exponents must be 0. $\boxed{2^0 = 3^0 = 1}$

$$4a - b - 6 = 0$$

$$b - 2a = 0$$

+--------------------------------

$$4a - b - 6 + b - 2a = 0$$

$$2a = 6$$

$$a = 3$$

$$b - 2a = 0$$

$$b = 6$$

So, $a + b =$

$$6 + 3 = 9$$

Review of Algebra and Linear Equations

Example 1.33 Binomial Expansions

If $x - \frac{1}{x} = 2\sqrt{2}$, what is the value of $\left(x + \frac{1}{x}\right)^2$?

Solution:

Let's take the square root of $x - \frac{1}{x} = 2\sqrt{2}$

$$\left(x - \frac{1}{x}\right)^2 = \left(2\sqrt{2}\right)^2$$

$$x^2 + \left(\frac{1}{x}\right)^2 - 2.x.\frac{1}{x} = 8$$

$$x^2 + \left(\frac{1}{x}\right)^2 = 10$$

$$\left(x + \frac{1}{x}\right)^2 = ?$$

$$x^2 + \left(\frac{1}{x}\right)^2 + 2.x.\frac{1}{x} = ?$$

We've already found $x^2 + \left(\frac{1}{x}\right)^2 = 10$

So; $\left(x + \frac{1}{x}\right)^2 = 10 + 2 = 12$

$$(a+b)^2 = a^2 + b^2 + 2ab$$
$$(a-b)^2 = a^2 + b^2 - 2ab$$
$$(a+b)^3 = a^3 + 3a^2b + 3ab^2 + b^3$$
$$(a-b)^3 = a^3 - 3a^2b + 3ab^2 - b^3$$

Example 1.34

If $140^2 - 137^2 = 3x$, what is the value of x?

Solution:

$$140^2 - 137^2 = 3x$$

$$(140 - 137)(140 + 137) = 3x$$

$$3(277) = 3x$$

$$a^2 - b^2 = (a-b)(a+b)$$
$$a^3 - b^3 = (a-b)(a^2 + ab + b^2)$$
$$a^3 + b^3 = (a+b)(a^2 - ab + b^2)$$

$x = 277$

Example 1.35

If: $366^2 - 363^2 = 3^x$, what is the value of x?

Solution:

$$366^2 - 363^2 = 3^x$$

$$(366 - 363)(366 + 363) = 3^x$$

$$3(729) = 3^x$$

$$3^1(3^6) = 3^x$$

$$3^{1+6} = 3^x$$

$$x = 7$$

> We need to factorize 729 by dividing it into 3, and keep dividing:
>
> $$\frac{729}{3} = 243; \quad \frac{243}{3} = 81; \quad \frac{81}{3} = 27$$
>
> $$\frac{27}{3} = 9; \quad \frac{9}{3} = 3; \quad \frac{3}{3} = 1$$

Example 1.36

If (x, y, z) satisfies the system of equations below, what is the value of $x^2 + z^2 - 2y^2$?

$$x - y = y - z = 4$$

Solution:

Since y appears to be the common variable, let's convert the other two variables (x and z) to y.

$$x = y + 4$$

$$z = y - 4$$

Now, let's put the new definitions into our value:

$$x^2 + z^2 - 2y^2 = ?$$

Review of Algebra and Linear Equations

$$(y+4)^2 + (y-4)^2 - 2y^2 = ?$$

$$y^2 + 8y + 16 + y^2 - 8y + 16 - 2y^2 = ?$$

$$2y^2 - 2y^2 + 8y - 8y + 32 = \mathbf{32}$$

Example 1.37

Find $\sqrt{\dfrac{49}{81} + \dfrac{1}{16} - \dfrac{7}{18}} = ?$ (without a calculator)

Solution:

$$\sqrt{\frac{7^2}{9^2} + \frac{1^2}{4^2} - \frac{7}{18}} =$$

$$\sqrt{\left(\frac{7}{9}\right)^2 + \left(\frac{1}{4}\right)^2 - 2 \cdot \left(\frac{7}{9}\right)\left(\frac{1}{4}\right)} =$$

$$\sqrt{\left(\frac{7}{9} - \frac{1}{4}\right)^2} =$$

$$\phantom{\sqrt{\Big(}}(4)(9)$$

$$\sqrt{\left(\frac{28-9}{36}\right)^2} = \frac{19}{36}$$

Example 1.38

If (a, b) satisfies the system of equation below, and of $a > 0$, and $b > 0$, what is the value of $\dfrac{a}{b}$?

$$\frac{a-b}{a\sqrt{b} + b\sqrt{a}} = \frac{1}{\sqrt{a}}$$

Solution:

$$\frac{a-b}{a\sqrt{b}+b\sqrt{a}} \cdot \frac{(a\sqrt{b}-b\sqrt{a})}{} = \frac{1}{\sqrt{a}}$$

$$\frac{(a-b)(a\sqrt{b}-b\sqrt{a})}{(a\sqrt{b}+b\sqrt{a})(a\sqrt{b}-b\sqrt{a})} = \frac{1}{\sqrt{a}}$$

$$\frac{(a-b)(a\sqrt{b}-b\sqrt{a})}{a^2b - b^2a} = \frac{1}{\sqrt{a}}$$

$$\frac{(a-b)(a\sqrt{b}-b\sqrt{a})}{ab(a-b)} = \frac{1}{\sqrt{a}}$$

$$\frac{(a\sqrt{b}-b\sqrt{a})}{ab} = \frac{1}{\sqrt{a}}$$

$$ab = \sqrt{a}(a\sqrt{b} - b\sqrt{a})$$

$$ab = a\sqrt{ab} - ba$$

$$2ab = a\sqrt{ab}$$

$$2b = \sqrt{ab}$$

$$4b^2 = ab$$

$$4b = a$$

$$\frac{a}{b} = 4$$

Example 1.39   Factoring

What is the solution set of the equation below?

$$x^2 + 2x - 8 = 0$$

Review of Algebra and Linear Equations

Solution:

$$x^2 + 2x - 8 = 0$$

| x | +4 |
| x | −2 |

$$(x + 4)(x - 2) = 0$$

$$x = -4$$
$$x = 2$$

Solution Set: {−4, 2}

Example 1.40

What is the solution set of the equation below?

$$x^2 - 7x + 12 = 0$$

| x | −3 |
| x | −4 |

$$(x - 3)(x - 4) = 0$$

$$x = 3$$
$$x = 4$$

Solution Set: {3, 4}

Example 1.41

What is the solution set of the equation below?

$$x^2 - 5x = 24$$

$$x^2 - 5x - 24 = 0$$

| x | +3 |
| x | −8 |

$(x + 3)(x − 8) = 0$

$x = −3$
$x = 8$

Solution Set: $\{−3, 8\}$

Example 1.42

What is the solution set of the equation below?

$$2x^2 + 21 = 17x$$

$2x^2 − 17x + 21 = 0$
2x   **−3**
x   **−7**

$(2x − 3)(x − 7) = 0$

$x = 3/2$
$x = 7$

Solution Set: $\{\frac{3}{2}, 7\}$

Example 1.43

As per the equation below find k in terms of one of the solutions of x?

$$kx^2 + 6kx + 3x = −18$$

$kx^2 + 6kx + 3x + 18 = 0$
kx   **+3**
x   **+6**

$(kx + 3)(x + 6) = 0$

$$x_1 = -6 \qquad x_2 = \frac{-3}{k}$$

So; $\quad k = \dfrac{-3}{x}$

Example 1.44

What is the solution set of the equation below?

$$x = \sqrt{4x - 4} + 1$$

Solution:

$$x - 1 = \sqrt{4x - 4}$$

$$(x - 1)^2 = (\sqrt{4x - 4})^2$$

$$x^2 - 2x + 1 = 4x - 4$$

$$x^2 - 6x + 5 = 0$$

| x | -5 |
| x | -1 |

$$(x - 5)(x - 1) = 0$$

$$x_1 = 5$$
$$x_2 = 1$$

Solution Set: {5, 1}

Example 1.45

How many solutions are there to the system of equations below?

$$2x^2 - xy = 6x$$

$$xy + 2x = x^3 - 2x^2$$

Solution:

Let us isolate the "xy".

$$xy = 2x^2 - 6x$$

$$xy = x^3 - 2x^2 - 2x$$

$$2x^2 - 6x = x^3 - 2x^2 - 2x$$

$$x^3 - 4x^2 + 4x = 0$$

$$x(x^2 - 4x + 4) = 0$$

$$x(x-2)(x-2) = 0$$

$$x_1 = 0$$
$$x_2 = 2$$

Solution Set: {0, 2} **There are 2 solutions.**

Example 1.46

What is the sum of the roots of the equation below?

$$(x^2 - 3x - 10)(x + 4) = 0$$

Solution:

$$(x^2 - 3x - 10)(x + 4) = 0$$
$$\begin{array}{cc} x & -5 \\ x & +2 \end{array}$$

$$(x - 5)(x + 2)(x + 4) = 0$$

Review of Algebra and Linear Equations

$$x = 5 \; ; \; x = -2 \; ; \; x = -4$$

Sum = $5 - 2 - 4 = -1$

Example 1.47

In the equation below, *a, b, c,* and *d* are constants. If the equation has roots $2, -3 \; and -1$, what is the value of $a + b + c + d$?

$$ax^3 + bx^2 + cx + d = 0$$

Solution:

Since we know the roots: $2, -3 \; and -1$; the equation will be:

$$(x - 2)(x + 3)(x + 1) = 0$$

Let's expand it:

$$(x - 2)(x^2 + 4x + 3) = 0$$

$$x^3 + 4x^2 + 3x - 2x^2 - 8x - 6 - 0$$

$$x^3 + 2x^2 - 5x - 6 = 0$$

$$x^3 + 2x^2 - 5x - 6 = ax^3 + bx^2 + cx + d = 0$$

So; $\quad a = 1 \; ; \; b = 2 \; ; \; c = -5 \; ; \; d = -6$

$$a + b + c + d = 1 + 2 - 5 - 6$$

$$a + b + c + d = -8$$

Example 1.48

If (*x, y, z, v*) satisfies the system of equations below, what is the value of $x + y + v$ in terms of *z*?

$$x + y + z = v$$

$$y + z + v = x$$

$$z + v + x = y$$

Solution:

We need to find a way to isolate $x + y + v$. Notice the right hand side of all three equations: *v*, *x*, *y*

So, let's add everything up:

$$2x + 2y + 2v + 3z = x + y + v$$

$$x + y + v + 3z = 0$$

$$x + y + v = -3z$$

Example 1.49 **Quadratic Equation and Quadratic Formula**

What is the solution set of the equation below?

$$2x^2 - 15 = 3x$$

$$2x^2 - 3x - 15 = 0$$

| **Warning!!** Factoring won't work here. |

When we cannot factorize an equation in the form of $ax^2 + bx + c = 0$; we can use the quadratic equation formula:

$$x = \frac{-b \pm \sqrt{b^2 - 4ac}}{2a}$$

| We can break this into two parts. |

$$x = \frac{-b + \sqrt{b^2 - 4ac}}{2a}, \text{ and } x = \frac{-b - \sqrt{b^2 - 4ac}}{2a}$$

Review of Algebra and Linear Equations

So; for our equation $2x^2 - 3x - 15 = 0$

a = 2

b = −3

c = −15

Warning!! Make sure the equation is in the same order as: $ax^2 + bx + c = 0$

Let's plug the numbers into their places:

$x = \dfrac{3+\sqrt{9+120}}{4}$, and $x = \dfrac{3-\sqrt{9+120}}{4}$

$x = \dfrac{3+\sqrt{129}}{4}$, and $x = \dfrac{3-\sqrt{129}}{4}$

Example 1.50

What is the product of the values of "x" that satisfy the equation below?

$x^2 - 4x = 4$

Solution:

$x^2 - 4x - 4 = 0$

Since factoring won't work here, let's use the quadratic equation formula.

$x = \dfrac{-b \pm \sqrt{b^2 - 4ac}}{2a}$

We can break this into two parts.

$x = \dfrac{-b + \sqrt{b^2 - 4ac}}{2a}$, and $x = \dfrac{-b - \sqrt{b^2 - 4ac}}{2a}$

So; for our equation $x^2 - 4x - 4 = 0$

a = 1

b = −4

c = −4

Warning!! Make sure the equation is in the same order as: $ax^2 + bx + c = 0$

Let's put the numbers in their places:

$$x = \frac{4+\sqrt{16+16}}{2}, \text{ and } x = \frac{4-\sqrt{16+16}}{2}$$

$$x = \frac{4+\sqrt{32}}{2}, \text{ and } x = \frac{4-\sqrt{32}}{2}$$

$$\boxed{\sqrt{32} = \sqrt{16(2)} = 4\sqrt{2}}$$

$$x = \frac{4+4\sqrt{2}}{2}, \text{ and } x = \frac{4-4\sqrt{2}}{2}$$

$x_1 = 2 + 2\sqrt{2}$, and $x_2 = 2 - 2\sqrt{2}$

Product of the values of x:

$P = (2 + 2\sqrt{2})(2 - 2\sqrt{2})$

$P = 4 - 8$

$P = -4$

To save time you can apply this formula:

$$x^2 - y^2 = (x-y)(x+y)$$

FYI! There is a formula to find the product of the roots of $ax^2 + bx + c = 0$

$$x_1 \cdot x_2 = \frac{c}{a}$$

Example 1.51

If k is a positive real number and is one of the roots of the equation below, what is the other root?

$$2x^2 + kx - 9 = 0$$

Solution:

The roots satisfy the equation. So, let's replace x with k.

$2k^2 + k \cdot k - 9 = 0$

$3k^2 = 9$

Review of Algebra and Linear Equations

$k^2 = 3$

$k = \pm\sqrt{3}$ We know that k is a positive real number.

$k = \sqrt{3}$

Now, let's apply the product of the roots formula:

$x_1 \cdot x_2 = \frac{c}{a}$ Since k is one of the roots; $k = x_1$

$\sqrt{3} \cdot x_2 = \frac{c}{a}$

$\sqrt{3} \cdot x_2 = \frac{-9}{2}$

$x_2 = \frac{-9}{2\sqrt{3}} = \frac{-9\sqrt{3}}{2 \cdot 3} = -\frac{3\sqrt{3}}{2}$

Practice Questions

Find the value of the missing variable:

1. $7x + 5 = 2x$

2. $3x - 1 + 2x + 5 = 4x + 6$

3. $x/2 + 3 + 2x = 17 - x$

4. $3x + 106 = 5x - 2$

5. $4x + x + x - 3 + x - 1 = 5 + x + x$

6. $m - (2m + 1) + 2m/2 = 3m - 2(5m + 1)/3$

7. $\{a - [b - (c + 2)]\} - \{a + [c - (a - 1)]\} + b - 1 = ?$

8. $\dfrac{1}{t+4} = \dfrac{3}{2t+5}$

9. $\dfrac{2x}{3} + 7 = 4x - 13$

10. $\dfrac{3y+2}{2} - \dfrac{y+4}{5} = 2y - 4$

11. $\dfrac{4m-5}{3} + 2 = \dfrac{3m-1}{2}$

12. $\dfrac{k-1}{k-3} = \dfrac{k-5}{k-4}$

13. $\dfrac{1-\frac{1}{a}}{1+\frac{1}{a}} = 3$

14. $\dfrac{\frac{2}{5}+\frac{3}{4}+\frac{3}{20}}{\frac{1}{2}+\frac{5}{3}} = \dfrac{3}{x}$

15. $\dfrac{4}{9-\frac{49}{9}} - \dfrac{1}{8} = x$

16. $9\left(\dfrac{1-3^{-4}}{1-3^{-2}}\right) = 5x$

17. $\left(1-\dfrac{4}{7}\right)\left(1-\dfrac{2}{9}\right)\left(1-\dfrac{6}{15}\right) = x^{-1}$

Review of Algebra and Linear Equations 33

18. $\dfrac{2^{-2}}{\frac{1}{x^{-1}}+4^{-1}} = 17^{-1}$

19. $\dfrac{3^2 \cdot 6^3 + 6^2 \cdot 3^3}{54} = 6x$

20. $\dfrac{6^{-2} - 4 \cdot 6^{-3}}{3^{-2} - 2 \cdot 3^{-3}} = x$

21. $9^x(12^{3-2x}) = 54$

22. $9^x \cdot 4^x \cdot 6^x = 36$

23. If $2^x = 6^{x+y-1}$, what is 3^x in terms of y?

24. $\sqrt[3]{\dfrac{32}{\sqrt{8}-\sqrt{2}}} = m$

25. $\dfrac{\sqrt[3]{2\sqrt{54}}}{\sqrt{2}} = x$

26. $x\left(\sqrt{\dfrac{1}{x}-\dfrac{1}{x^2}}\right) = 2$

27. $\sqrt{\dfrac{\sqrt{x}+3}{\sqrt{x}-3}} = \sqrt{x}+3$

28. $\dfrac{1+\sqrt{x}}{1-x} - \dfrac{x}{1-\sqrt{x}} = \dfrac{7}{2}$

29. $\dfrac{1}{\sqrt{2a}} + \dfrac{4}{\sqrt{8a}} = 3$

30. $\dfrac{1}{a} + \dfrac{a}{a+1} + \dfrac{a-1}{a} = \dfrac{4}{3}$

31. $(1 - 3^{-1} + x^{-1})^{-3} = 64$

32. If $x \neq 0, y \neq 0, z \neq 0$, and $x + y + z = xy$; what is the simplified form of:

$$\frac{xy + xz + yz + z^2}{xyz}$$

33. If $2x + 3 - \frac{2x^2 + 3x - 9}{2x - 3} = m$, what is m in terms of x?

34. If a, b, and c are natural numbers that satisfy the system of equations below, what is the value of $a+b$?

$$\frac{a}{b} = \frac{b}{c} \; ; \; a^2 + ac + 2ab = 16$$

35. What is the solution set of the equation below?
$x^2 - 6x + 5 = 0$

36. What is the solution set of the equation below?
$6x^2 - 11x = 10$

37. What is the solution set of the equation below?
$2x^2 - 4x = 30$

38. What is the solution set of the equation below?
$2x^2 - 3x\sqrt{5} + 5 = 0$

39. What is the solution set of the equation below?
$\sqrt{14 - x} = x - 2$

40. When $t \neq 0$, find the product of the solutions of the equation below.

$2ty^2 + ty - 10t = 0$

Review of Algebra and Linear Equations

41. What is the sum of the values of x that satisfy the system of equation below?

    $2x^2 + 5 = 8x$

42. What is the sum of the values of x that satisfy the system of equation below?
    $(2x - 3)(x + 1) + (x - 2)(2x - 3) = 0$

43. Simplify: $\dfrac{a}{a-1} + \dfrac{a}{a+1} - 2$

44. Simplify: $\dfrac{3kx^2 - 6k^2x}{2kx^3 - 8k^3x}$

45. Simplify: $\dfrac{x^3 + y^3}{(x-y)^2 + xy}$

46. Simplify: $\dfrac{1 + mnx^2 - (m+n)x}{mx - 1}$

47. Simplify: $\dfrac{x^{-1} + x^2}{x + x^{-2}}$

48. Simplify: $\dfrac{x^{-2} + x}{x^{-2} - x^{-1} + 1}$

49. Simplify: $\left[\dfrac{a}{1+\frac{a}{b}} - \dfrac{b}{1-\frac{b}{a}} \right] \div \dfrac{ab}{(a^2 - b^2)}$

50. If (a, b) satisfies the system of equations below, what is the value of (a + b) ?

    $3ab^2 + a^3 = 9$
    $3a^2b + b^3 = 18$

51. The sum of $4x^3 + 2x^2 - 5x - 2$ and $3x^2 + 2x + 4$ can be written in the form $kx^3 + hx^2 + mx + n$; where k, h, m and n are constants. What is the value of $\dfrac{k+h}{m+n}$?

52. If x is an integer which satisfies $(-4 < x < 3)$, what is the minimum value of $(2-x)(2+x)$?

53. In the equation $(kx - 3)^2 = 81$, k is a constant. If $x = -6$ is one solution to the equation, what is a possible value of k?
a) -6 b) -2 c) 0   d) 1   e) 2   f) 6

54. If $x \neq 0$; what is the sum of the values of x that satisfy the system of equation below?
$$1 + \dfrac{2}{x} - \dfrac{3}{x^2} = 0$$

55. If (a, b) satisfies the system of equations below, what is the value of b?
$$a - 4 = b$$
$$2b + 3a = 7$$

56. What is the ordered pair (x, y) that satisfies the system of equations below?
$$2x = 1 - y$$
$$\dfrac{x}{3} + \dfrac{y}{2} = \dfrac{y-x}{6}$$

57. If (x, y) satisfies the system of equations below, what is the value of x?
$$2y = 4x + 12$$
$$3x + 3y = -18$$

58. The expression below can be written in the form $kx^2 + h$, where k and h are constants. What is the value of $k - h$?
$$12(20x^2 - 40) - 5(20 - 3x^2)$$

59. If (h, k, r) satisfies the system of equations below, what is the ratio of h to k?

$$\frac{h+r}{k} = \frac{3}{2}$$

$$\frac{k}{r} = \frac{3}{4}$$

60. If (h, k, r, x) satisfies the system of equations below, what is x?

$$\frac{k}{x} + r = \frac{k(h+r)}{hx}$$

61. If (x, y) satisfies the system of equation below, what is the value of $(x+y)^{-1}$?

$$\frac{1}{8} + 3x = \frac{1}{2} - 3y$$

62. What is the value of: $x^2 + xy - z(x+y)$; if:

$x + y = 5$

$x - z = -1$

63. Find $\left(x - \frac{1}{x}\right)^2$, when $x + \frac{1}{x} = 4\sqrt{2}$

64. If (a, b) satisfies the system of equations below, what is the value of a + b?

$$a^2 - b^2 = 10 \quad ; \quad \frac{4^{a-b}}{4^{b-a}} = 16$$

65. If (a, b) satisfies the system of equations below, and b < 0; what is the value of b?

$$a^2 - b^2 = 27$$

$$\frac{1}{a-b} + \frac{1}{a+b} = \frac{4}{9}$$

66. If $x = 3^{1/4} + 1$, find $\dfrac{(3^{1/8}+1)(3^{1/8}-1)}{\sqrt{3}-1}$ in terms of x.

67. If $a = 10/3$, find $(a-5)^3 + 3(a-5)^2 + 3(a-5) + 1$.

68. If (a, b) satisfies the system of equations below, what is the value of a?
$a^4 - 2ab = 21$
$a^3 - 2b = 7$

69. Simplify: $\left(\frac{1}{a} + \frac{1}{b}\right)\left(\frac{1}{a} - \frac{1}{b}\right) + \left(\frac{2}{a} + \frac{2}{b}\right)\left(\frac{2}{a} - \frac{2}{b}\right)$

70. Simplify: $\frac{4x^3+16x^2}{4x^2+12x} \div \frac{x^3-16x}{x^2-x-12}$

71. Find $\frac{256a^4+1}{16a^2}$, if $4a + \frac{1}{4a} = 7$

72. If $x < y$, simplify: $\sqrt{\frac{5^x}{5^{-y}}\left(-2 + \frac{5^x}{5^y} + \frac{5^y}{5^x}\right)}$

73. $a = x^{1/3} + y^{2/3}$; and $b = x^{1/3} - y^{2/3}$, what is $(a^2 - b^2)^3$ in terms of x and y?

74. If a and b are natural numbers that satisfy the system of equation below, and
$a < 5$; $b < 5$; what is the value of $(a + b)$?
$\frac{260}{100} = a + \frac{b}{5}$

75. If $x - \frac{1}{x} = 3\sqrt{2}$, what is the value of $\left(x + \frac{1}{x}\right)^2$?

76. If (a, b) satisfies the system of equations below, what is the value of $a^2 + b^2$?

$a^2 - a = b^2 - b$
$ab = -1$

Review of Algebra and Linear Equations

77. If x is a real number that satisfies the system of equation below, and k is a positive integer, what is the value of k?
$$\frac{(x+3)^2 - 2x(x+3) + x^2}{(7-x) - (k-x)} = 3$$

78. If (a, b) satisfies the system of equation below, what is the value of $\frac{b}{a}$?
$$\frac{2b}{a+\frac{1}{b}} - \frac{3a}{b+\frac{1}{a}} = \frac{5a^2}{ab+1}$$

79. If (a, b) satisfies the system of equation below, what is the product of a and b?
$$\sqrt{a} - \sqrt{b} = \sqrt{a+b-1}$$

80. If (a, b) satisfies the system of equations below, what is the product of a and b?
$$a = \sqrt{12} - \sqrt{8}$$
$$b = \sqrt{27} + \sqrt{18}$$

81. If (a, b) satisfies the system of equations below, what is the product of a and b?
$$b = 2 - a$$
$$\frac{1}{a} + \frac{1}{b} = \frac{1}{4}$$

82. If (x, y) satisfies the system of equations below; $x > 0$; and $y > 0$, what is the value of $x + y$?
$$x^2 + y^2 = 12$$
$$\frac{1}{x} + \frac{1}{y} = 2$$

83. If (a, b, c) satisfies the system of equation below, what is the value of $(a + b + c)^2$?
$$4a + 5b + 6c = 14 \quad ; \quad a + 2b + 3c = 5$$

84. If (a, b, c) satisfies the system of equation below, what is the value of a?
    $2a + 2b - c = 1$
    $a + b + c = 2$
    $b - c = 1$

85. If (x, a, b, c) satisfies the system of equation below, what is the value of $a.b.c$?
    $2x - 4 = ax(x - 1) + bx(x + 1) + c(x^2 - 1)$

86. If (x, y) satisfies the system of equations below, what is the value of $\frac{x+y-1}{xy}$?
    $x = \frac{a}{a-b}$
    $y = \frac{b}{a+b}$

87. If a, b, x, y are positive real numbers which satisfy the system of equations below, what is x in terms of a?
    $\frac{x}{a} \cdot \frac{b}{y} = 2$
    $\frac{a^2}{x^2} + \frac{b^2}{y^2} = 20$

88. If (x, y, z) satisfies the system of equations below, what is the value of x?
    $x \cdot y = 14$
    $x \cdot z = 20$
    $3x + 2y + z = 24$

89. If $a \geq 0, b \geq 0, c \geq 0$; and a, b, c are integers which satisfy the system of equations below, and $a \neq b \neq c$; what can be maximum value of c?
    $3a + 2b + c = 60$
    $2a + 3b + c = 50$

Review of Algebra and Linear Equations

90. If (x, y, z, v) satisfies the system of equations below, what is the value of $x - 2y - 2z + v$?

    $x - y = 25$
    $y + z = 13$
    $z - v = 10$

91. If a, b, c are positive integers which satisfy the system of equations below and $(a \neq b \neq c)$; what is the sum of the values that a can get?

    $a - b + c = 3$
    $a + b + c = 7$

92. If the product and the sum of the roots of the equation below is -6 and -5, respectively; what is the value of b?

    $ax^2 + bx = 12$

93. When 2^{645} is multiplied out, what number appears in the units digit?

94. When 3^{256} is multiplied out, what number appears in the units digit?

CHAPTER II

Problem Solving

Example 2.1   Basic Short Problems

One pound of apples costs $3. How much will *y* pounds of apples cost at this rate?

Solution:

If   1 pound......... $3
         y pounds....... ? $(*x*)

As the pounds increase, so will the price of apples. This is a directly proportionate relationship.

When there is a directly proportionate relationship, we do cross multiplication.

$y(3) = 1(x)$

$x = 3y$

$x = \$3y$

Example 2.2

Patrick is a skateboarder and can travel at a speed of 15 miles per hour. How fast can Patrick go in feet per minute? (1 mile = 5280 feet)

Solution:

Let's convert "15 miles per hour" to minutes:

15 miles per hour = 15 miles per 60 minutes

If   60 minutes .... 15 miles

         1 minute .... x miles

$$x = 15/60$$

$$x = 0.25 \, miles$$

Finally, let's convert miles to feet:

If   1 mile .... 5280 feet

         0.25 .... y feet

$$y = 5280(0.25)/1$$

$$\mathbf{y = 1320 \, feet}$$

Example 2.3

Out of 125 apples in a factory, 20 are rotten. How many apples can be assumed rotten at the same rate if there are 500 apples?

Solution:

If   125 apples ..... 20 rotten
           500 apples ..... x rotten

As the number of apples increase, so will the number of rotten apples. This is a directly proportionate relationship.

Problem Solving

When there is a directly proportionate relationship, we do cross multiplication.

$$500(20) = 125(x)$$

$$x = \frac{10,000}{125} = 80 \; rotten \; apples$$

Example 2.4

In a furniture factory 120 workers finish upholstering 300 sofa beds in 2 days. Due to economic recession, 40 workers get laid off. How many days will it take for the remaining workers to get 300 sofa beds upholstered?

Solution:

Remaining # of workers: 120 − 40 = 80

If 120 workers ……. 2 days
    80 workers …….. x days

In this situation we have an inversely proportionate relationship because as the number of workers decrease, the number of days for the workers to finish the job will increase. (Apply common sense)

So, $120(2) = 80(x)$

$$x = \frac{240}{80} = 3 \; days$$

Warning!! Instead of cross multiplication, we do adjacent multiplication.

Example 2.5

In the same furniture factory we know that now 80 workers finish upholstering 300 sofa beds in 3 days. All of a sudden economy picks up and the demand reaches 400 beds. Top management hires 20 new workers to meet the demand. How long will it take for the total workers to get 400 sofa beds upholstered?

Solution:

Total workers: 80 + 20 = 100

If   80 workers ... 3 days ...... 300 sofas
      100 workers ...x days .... 400 sofas

We better break this three-way relationship into 2 parts.

1. part: 80 workers ... 3 days ...... 300 sofas
         100 workers ...x days .... 300 sofas

Warning!! We assume 100 workers are also finishing 300 sofas.

This first relationship is inversely proportional. We can exclude the assumption which is finishing 300 sofas.

So; $80(3) = 100(x)$

$$x = \frac{240}{100} = 2.4 \; days$$

Now, we know that it takes 2.4 days for 100 workers to finish 300 sofas. How long will it take for them to finish 400 sofas?

2. part   2.4 days ..... 300 sofas
          y days .... 400 sofas

This second relationship is directly proportional.

So; $300(y) = 400(2.4)$

$$y = \frac{960}{300} = 3.2 \; days$$

If   1 day ..... 24 hours
      3.2 days ... t hours

If necessary we can do more...

This day-to-hour conversion relationship is also directly proportional.

So; $24(3.2) = 1(t)$

$$t = 76.8 \; hours$$

Problem Solving

Example 2.6   Formula-based Problems

A new tech company started-up in a garage with 5 software engineers. The company plans to hire 4 more software engineers every quarter (1 quarter is 3 months) for the next 2 years. If the total number of software engineers "e" can be represented with an equation in the form of:

$e = hx + k$, where "x" is the number of quarters that have passed after the opening of the company, what is the value of k?

Solution:

$e = hx + k$

If "x" is the number of quarters, then "h" is 4, because the company plans to hire 4 more software engineers every quarter.

When the company first started-up, there are 0 quarters, yet there are 5 engineers.

So, $e = k = 5$

Example 2.7

Refer back to the previous example and calculate the total number of software engineers 2 years after the company started-up.

Solution:

The company plans to hire 4 more software engineers every quarter. So; $h = 4$

We know that $k = 5$. So; our equation looks like this:

$e = 4x + 5$

Let's read the question once again. It says: "x" is the number of quarters that have passed after the opening of the company. We want to find out after 2 years which is equivalent of 8 quarters (1 year has 4 quarters, so 2 years have 8 quarters). So, x = 8.

$e = 4.8 + 5$

$e = 37$

> **Warning!!** We may use period (.) as a sign of multiplication from time to time.

Example 2.8

A group of 156 college students went on a summer holiday. They booked 105 rooms in total. Some of them stayed in double (2 students per room) and some stayed in single (1 student per room) rooms. How many of the students stayed in single rooms?

Solution:

Let's say   number of double rooms = y
             number of single rooms = x

Since 105 rooms were booked;

$$x + y = 105$$

Now we need to record the number of students in terms of the number of rooms. Since we have 156 students;

$$x(1) + y(2) = 156$$

We can solve the problem by using these equations.

$$x + 2y = 156$$
$$x + y = 105$$
$$-\;\dotfill$$
$$x + 2y - (x + y) = 156 - 105$$

$$x + 2y - x - y = 51$$

$y = 51$ (# of double rooms)

$x = 105 - 51$

x = 54 (# of single rooms)

Problem Solving

Example 2.9

Refer back to the example above and calculate the number of students that stayed in double rooms.

Solution:

We know that they booked 51 double rooms. So;

51(2) = **102 students**

Example 2.10

A food truck sells wraps for $7.00 each and drinks for $2.00 each. The food truck's revenue from selling a total of 237 wraps and drinks in one day was $984. How many drinks were sold that day?

Solution:

# of wraps sold: x , selling price= 7

# of drinks sold: y , selling price=2 , $y = ?$

$$7x + 2y = 984$$

$$x + y = 237$$

$$7x + 2y = 984$$

$$-2\,(x + y = 237) - 2$$

| **Tip!** We want to get rid of one of the variables. We usually prefer the easier variable. |

$$7x - 2x + 2y - 2y = 984 - 474$$

$$5x = 510$$

$$x = 102 \ (\text{\# of wraps sold})$$

$e = 4.8 + 5$

$x + y = 237$

$y = 135$ (# of drinks sold)

Example 2.11 Percentage Problems

20% of a number is 15 less than half of the same number. What is the number?

Solution:

$$x\left(\frac{20}{100}\right) = \frac{x}{2} - 15$$

Let's simplify.

$$\frac{x}{5} = \frac{x}{2} - 15$$

$$\frac{x}{2}_{(5)} - \frac{x}{5}_{(2)} = 15$$

$$\frac{5x - 2x}{10} = 15$$

$$\frac{3x}{10} = 15$$

$$3x = 150$$

$$x = 50$$

Example 2.12

A recent lab analysis indicated that the water produced at your bottling plant contains 2.2 mg/L Chloride. By what percent should you reduce the Chloride to attain 1.2 mg/L level?

Solution:

Problem Solving 51

Existing Chloride level : 2.2 mg/L

Desired Chloride level : 1.2 mg/L

By how much should you reduce it to reach the desired level?

2.2 − 1.2 = 1 mg/L

We need to find the percentage of the difference of 1 mg/L applicable to the existing level.

If 2.2 .... 100 %
1 .... x %

$x = 100/2.2$

$x = 45.45\%$

Example 2.13

A shop owner has a profit margin of 30% on everything she sells in her shop. On last Boxing Day she offered a special discount of 20% on shoes only. How much profit or loss did she make on the shoes that she sold on that day?

Solution:

Let us assume one pair of shoes costs $100.

30% more is:

100(1.3) = 130   or   100 + 100(0.3) = 130

After 20% discount, the selling price will be:

130(0.8) = 104   or   130 − 130(0.2) = 104

The difference between the initial cost ($100) and the final selling price ($104) is either profit or loss.

104 - 100 = 4 (Since the selling price is more than the initial cost, it is profit)

Finally, we need to find the percentage of this profit. What percent of 100 is 4?

$100(\frac{x}{100}) = 4$

x = 4% profit

Example 2.14

A grocery sold 40% of its fruits and vegetables with 60% profit, and 60% of its fruits and vegetables with 30% loss. What is the grocery's percent profit/loss?

Solution:

We can assume the grocery has 100 units of fruits & veggies, and each item costs $100.

So, the total cost of all items is $= 100 \cdot 100 = \$10{,}000$

40% of the items $= 100 \cdot \frac{40}{100} = 40$ items

60% profit $= 100 + 100 \cdot \frac{60}{100} = \$160$ (per item); or $= 100(1.6) = \$160$

60% of the items $= 100 \cdot \frac{60}{100} = 60$ items

30% loss $= 100 - 100 \cdot \frac{30}{100} = \$70$ (per item); or $= 100(0.7) = \$70$

....

Now, let's find how much sales the grocery made:

Sales $= 40(160) + 60(70)$

Sales $= \$6400 + \$4200$

Problem Solving

Sales = $10,600

Since Sales ($10,600) is more than costs ($10,000), we have profit. Let's calculate its percentage:

Profit = $10,600 − $10,000

Profit = $600

Profit Percent (x) = $10,000 $\cdot \frac{x}{100}$ = $600

Profit Percent (x) = 6%

> **Warning!** The profit percent must be based on the initial amount which is the cost ($10,000).

Example 2.15

Toni's small business made 30% more profit this year than last year. His business made last year 20% more profit than the year before. If Toni's business made $50x profit the year before, how much did it make this year in terms of "x"?

Solution:

The year before : 50x

Last year   : 20% more than the year before

$\qquad$ = 50x + 50x$(\frac{20}{100})$

$\qquad$ = 50x + 10x

$\qquad$ = 60x

This year   : 30% more than last year

$\qquad$ = 60x + 60x$(\frac{30}{100})$

$\qquad$ = 60x + 18x

$\qquad$ **= 78x**

Example 2.16

A second-hand furniture warehouse sold initially 10% of its inventory in January, then 20% of what was left from January in February, and then 25% of what was left from February in March. If the remaining inventory at the end of March was 10,800 units, how many units were initially in the warehouse at the beginning of the year?

Solution:

Let x be the number of units in the warehouse at the beginning of the year:

January: 10% of inventory sold: $(0.9x)$

February: 20% of inventory from January: $(0.9x)(0.8)$

March: 25% of inventory from February: $(0.9x)(0.8)(0.75)$

So; $10,800 = (0.9x)(0.8)(0.75)$

$10,800 = (0.72x)\frac{75}{100}$

$10,800 = (0.72x)\frac{3}{4}$

$10,800 = (0.18x)3$

$3,600 = 0.18x$

$360,000 = 18x$

$20,000 = x$

> Let's do the math without using a calculator!

Example 2.17

The ratio of average household income to average household expenditure in Country A is 1.25. In Country B the average household income is 10% less and average household expenditure is 10% more than those of Country A.

Problem Solving

What is the ratio of average household income to average household expenditure in Country B?

Solution:

For Country A: $\dfrac{average\ household\ income}{average\ household\ expenditure} = \dfrac{i}{e} = 1.25$

We can assign easy-to-calculate numbers to Country A, such as:

$\dfrac{125}{100} = \dfrac{i}{e} = 1.25$

For Country B : $\dfrac{10\%\ less\ than\ Country\ A}{10\%\ more\ than\ Country\ A} = \dfrac{125(0.9)}{100(1.1)} = \mathbf{1.022}$

Alternative Solution:

For Country A: $\dfrac{average\ household\ income}{average\ household\ expenditure} = \dfrac{i}{e} = 1.25$

For Country A: $i = 1.25e$

For Country B: $\dfrac{10\%\ less\ than\ Country\ A}{10\%\ more\ than\ Country\ A} = \dfrac{1.25e(0.9)}{e(1.1)} = \mathbf{1.022}$

Example 2.18

How much water (in gallons) must be vaporized to increase the salt percentage of a 50 gallon salt-water mixture from 10% to 20%?

Solution:

The current condition is 10% salt in the mixture. Since we will vaporize water, the amount of the salt must stay the same. So, the amount of salt:

Salt Amount = $50(0.1) = 5 \; gallons$

Let's say *x* amount of water will be vaporized; the amount of salt will be 20% of the new mixture:

5 gallons of salt = $(50 - x)(0.2)$

$5\left(\frac{10}{2}\right) = 50 - x$

$25 = 50 - x$

$x = 25 \; gallons$

Example 2.19

The value of a brand new computer drops 20% every 6 months after it gets sold and its value keeps decreasing at the same rate. If the initial sale price of one of these computers is $1200, what is its second-hand value after 2 years?

Solution:

Let us find the value of the computer after the first 6 months:

1200(0.8) = 960   or   1200 − 1200(0.2) = 960

In 2 years there are four 6 months and since the value keeps dropping at the same rate, 0.8 decrease will apply four times.

(((1200(0.8))0.8)0.8)0.8 = 491.5   or $1200(0.8)^4 = 491.5$
    6mo   1y 1.5y 2 y

Example 2.20

The number of female passengers booking a particular airline company's tickets is less than 40% of the male passengers booking the same company's tickets. If 601 female passengers booked a flight over the last week, calculate at least how many male passengers booked a flight over the last week.

Problem Solving

Solution:

f : the number of female passengers
m: the number of male passengers

$$f < m(\frac{40}{100})$$

$$f < m(\frac{2}{5})$$

$$f < \frac{2m}{5}$$

We know the number of female passengers; 601:

$$601 < \frac{2m}{5}$$

$$3005 < 2m$$

$$1502.5 < m$$

$$m > 1502.5$$

So, *m* can be at least **1503** passengers

Example 2.21

A green energy power plant produces on average $34million worth of energy every year. The plant management plans to spend $120million on installing an additional wind power generator which is expected to increase the current production value by 25%. How long will it take for the new wind power generator investment to pay off?

Solution:

For the new investment to pay off, the increase in the production value must break even $120million.

34,000,000(0.25) = 8,500,000 is the increase in the production value every year.

$\frac{120,000,000}{8,500,000}$ = 14.11 years = 14 years + 0.11 years = 14 + 365(0.11) = 14 years and 41 days

Example 2.22   Compound Interest

Answer the following questions based on the compound interest formula:

$$A = P\left(1 + \frac{r}{n}\right)^{nt}$$

where;

A: Amount accumulated
P: Principal (initial investment amount)
r: interest rate per year (decimal)
n: number of compounding times per year
t: number of years

Q_1: If $5000 is invested in a savings account that earns interest at the rate of 3.5% compounded quarterly, what is the value of the account at the end of 3 years?

Solution:

$A = P\left(1 + \frac{r}{n}\right)^{nt}$

$A = 5000\left(1 + \frac{0.035}{4}\right)^{4(3)}$

$A = 5000(1.00875)^{12} = \$5,551$

> Quarterly compounding means that money gets compounded 4 times a year. (n=4)

> In your calculator, use x^y function to solve large exponential calculations.

Problem Solving

Q₂: As an alternative to the quarterly compounding scenario above, the investor decides to compound the account monthly. In comparison to the quarterly scenario above, what will be the difference in the value of the account when it gets compounded monthly at the same rate after 3 years?

$$A = P\left(1 + \frac{r}{n}\right)^{nt}$$

$$A = 5000\left(1 + \frac{0.035}{12}\right)^{12(3)}$$

Monthly compounding means that money gets compounded 12 times a year. (n=12)

$$A = 5000(1.002916)^{36} = \$5{,}552.7$$

In your calculator, use x^y function to solve large exponential calculations.

So, the difference is $5{,}551 - 5{,}552.7 = \$1.7$

Q₃: A certificate of deposit was bought for $12,500 at the end of 2006 and was sold at the end of 2017. If the certificate earned 3% compounded monthly, what was its value at the end of 2017?

Solution:

$$A = P\left(1 + \frac{r}{n}\right)^{nt}$$

$$A = 12{,}500\left(1 + \frac{0.03}{12}\right)^{12(2017-2006)}$$

Monthly compounding means that money gets compounded 12 times a year (n=12). The number of years, t, is found by $2017 - 2006 = 11$

$$A = 12{,}500(1.002916)^{132} = \$17{,}380$$

Practice Questions:

1) 30% of a number is 8 less than half of the same number. What is the number?

2) 30% of a number is 3 more than one quarter of the same number. What is the number?

3) The price of a kettle is reduced by 15 percent. If the new price is then further reduced by 20%, what percent of a single reduction is equal to the two reductions together?

4) The forecasted population, P, of a residential area is given by:
$$P = 12,520(e)^{\frac{0.012n}{n^2+0.78}}$$
where n is the number of years after 1998. What was the population of this residential area in 1998?

5) In a class of 65 linguistics students, 32 are taking Spanish course, 30 are taking Italian course. Of the students taking Spanish and Italian, 7 are taking both courses. How many students are not enrolled in either course?

6) Ray's small business made 30% more profit this year than last year. His business made last year 20% more profit than the year before. If Ray's business made $x profit the year before, how much did it make this year in terms of "x"?

7) Jane works in a law firm and spends 22% of her 10-hour workday talking with clients on the phone. How many hours and minutes does she spend talking on the phone at work with clients?
a) 2 hours and 22 minutes   c) 2 hours and 12 minutes
b) 1 hour and 58 minutes   d) 2 hours and 2 minutes

Problem Solving

8) When a rubber band gets fully extended its original length increases by 120%. If the fully extended length of the rubber band is 660 meters, what is its original length?

9) In a law firm' office the Inkjet Printer prints out 2 pages per second and the Lazer Printer prints out 3 pages per second. One of the legal associates starts using both printers at the same time. After a while he notices that once the Inkjet Printer finished printing out the 60th page, the Lazer Printer has 30 more pages to finish its job. If both printers completed their jobs at the same time, how many pages in total has the associate printed out with these printers?

10) A marine animal at a theme park consumes 39 pounds of fish in six days. If it continues to get fed at the same rate, in how many days will its total consumption reach 91 pounds?

11) Dorothy has a scheduled meeting in her workplace at 9:00am sharp. Her house is 1 hour walking distance away from where she works. She started walking at a constant pace so as to be just on time at work for the meeting. Exactly halfway she realized that she had left some important presentation documents home. She ran back home, picked up the documents and kept running back to work at the same pace and speed. As a result she arrived in her workplace exactly at the scheduled time of the meeting. If Megan used the same route between home and work, what was the time when she picked up the documents?

12) In a job placement test, a candidate earns 40 points for every correct answer and loses 50 points for every wrong answer. If one of the candidates answered 30 questions and scored 300 points in total, how many questions would he have answered incorrectly?

13) It takes a track runner 21 minutes to run around a race course which is 6 miles long. It takes him 24 minutes to run the same distance the second time. What is his average speed in miles per hours for both runs?

14) Gili's bakery charges its customers $100 for 40 muffins and 50 bagels. Some party planner orders 30 muffins and 50 bagels, gives $100 to the bakery and receives $A as change. At Gili's bakery what is the selling price of 1 muffin and 1 bagel in terms of A?

15) A skirt in a retail shop is marked at $T. During the discount sale its price gets reduced by 25%. Staff members are allowed a further discount on the already discounted price. If a member of the staff purchases the skirt, what will she have to pay in terms of T?

16) An event organizer buys sound equipment for $22,860. The equipment depreciates in value at a constant rate for 9 years, after which it is considered to have no financial value. How much is the sound equipment worth 3 years after it is bought?

17) A party planner purchases a minivan for $30,000. The minivan depreciates in value at a rate of x% per annum. If 5 years after the purchase the minivan is worth $17,714.7; what is the annual depreciation rate (x%) of the minivan? (use a calculator)

18) If you invest $20K at 7.5% interest for 8 years in a row, what will be your total money (approximately) after 8 years?

19) If you invest $50K at 6.5% nominal (annual) interest compounded quarterly for 3 months, how much interest (approximately) will you earn after 3 months?

20) If you invest $70K at 4.5% nominal (annual) interest compounded monthly for 5 months, how much interest (approximately) will you earn after 5 months?

21) Almost how many years will it take for $9500 to amount to $21,000 at an annual interest rate of 5% compounded quarterly?

Problem Solving

22) When Bryan gets a salary raise which is half of Daniel's salary, the sum of Bryan's and Daniel's current salaries will equal to twice the initial salary of Bryan (B). What is Daniel's (D) salary in terms of Bryan's (B) initial salary?

23) In a country with 20% inflation rate in 2017, an engineer makes $50,000 a year. At least how much should the same engineer earn a year to continue having the same standard of living in 2019?

24) The price of 1 liter (lt.) of olive oil was $P. Due to shortage of supply and increased demand, the price of all olive oil products were raised by 25%. How much olive oil can a customer buy with $P in his pocket?
    a)1.2P   b)2.2lt.   c)0.6P   d)0.8 lt.   e)1.2P   f)0.8P

25) A gas station had a profit margin of 10%. If the cashier collected $5500 worth of sales, however $1000 cash in the register turned out to be fake, what would be the percentage loss of the gas station?

26) A second-hand car dealer buys a sports car at a price which is 30% less than its base auction price. If the car gets sold at the auction at a price which is 40% more than its base auction price, what is the dealer's percent profit in this transaction?

27) A college student saves 10% of her weekly stipend. If her savings reach $1040 exactly 1 year after she started saving, how much is her weekly stipend?

28) A group of friends are rent-sharing a house. They equally contribute to the weekly rent of $120. When one more person joins them the rent paid per person reduces by 25%. How much rent does each person pay per week in the recent situation?

29) In a class the number of girls is 40% of the number of boys. If the number of girls is more than 20, at least how many boys are there in the class?

30) If m is 16% of n and 25% of k is n, then what percent of k is m?

31) A trader makes 10% profit selling an item for $99, whereas he makes 10% loss selling another item for $99. What is the trader's loss/profit situation?
    a) Break-even (no loss, no profit)
    b) 4% profit
    c) 2% loss
    d) 4% loss
    e) 2% profit
    f) 6% loss
    g) 3% profit

32) When a sports shop increases the price of its items by 20%, the managers notice a 20% drop in the number of items sold in the shop. What is the change in the shop's revenue? (Revenue is the money made selling items)

33) There are 40 male students in a class. 70% of the female students and 32 male students have passed the mathematics course. If 75% of all the students in total have passed the mathematics course, what is the total class population?

34) An office manager has a monthly salary of $$k$ in January 2018. Three months later due to cutbacks her salary gets reduced by 30% and by November when economy picks up, she gets a raise by 30%. What is the difference between her salaries in January and November in terms of k?

35) How much water (in gallons) must be vaporized to increase the salt percentage of a 40 gallon salt-water mixture from 20% to 25%?

Problem Solving

36) Suzy is working this summer as part of an entertainment team in a holiday village. She earned $9 per hour for the first 15 hours she worked this week. Because of her outstanding performance the team leader decided to raise her salary to $12 per hour for the rest of the week. Suzy saves 80% of her earnings from each week. What is the least number of hours Suzy must work for the rest of the week to save at least $300 for the week?

37) In a manufacturing plant x, y, z amounts of output are produced by three different brands of equipment. If the output capacity of the equipment which produces x amount of output is increased by 20%, and the other two equipments' capacity gets reduced by 5%, the total output remains unchanged. What is the relationship between x, y, and z?

38) A tech store sells two types of supercomputers: C1 and C2. The price of C1 and C2 together is $75,000. C1 is twice as expensive as C2. The store starts to offer 10% and 20% discount on C1 and C2, respectively if both computers are purchased at the same time by a customer. What is the combined percent reduction in the sale of both computers when purchased by a customer at the same time?

39) In a lab experiment male mice and female mice were given 1 tablet every 12 hours and 8 hours, respectively. The tablets that were given to male mice contained 0.5 grams of proguanil each, whereas the tablets that were given to female mice contained 1 gram of proguanil each. If 95 tablets that contained 85 grams of proguanil were given to the mice altogether, how many mice were used in this experiment?

CHAPTER III

Chart, Table and Data Analysis

Example 3.1

Dosage Information for Laboratory Mice

	High Dosage	Low Dosage	TOTAL
Male Mice	20	25	45
Female Mice	22	10	32
TOTAL	42	35	77

Question 1) The table above demonstrates the level of a new experimental drug dosage that was given to lab mice. What fraction of the female mice were given low dosage?

Answer:

$(Number\ of\ Female\ Mice).(fraction) = Low\ Dosage\ for\ Female\ Mice$

$$22.f = 10$$

$$f = \frac{10}{22}$$

Question 2) What fraction of the mice were given low dosage?

Answer:

$(Number\ of\ All\ Mice).(fraction) = Low\ Dosage\ for\ All\ Mice$

$$77.f = 35$$

$$f = \frac{35}{77}$$

Question 3) What percent of the male mice were given high dosage?

Answer:

$$(Number\ of\ Male\ Mice).(percent) = High\ Dosage\ for\ Male\ Mice$$

$$45 \cdot \frac{p}{100} = 20$$

$$p = \frac{20 \cdot 100}{45} = \frac{400}{9} = 44.44\ \%$$

Example 3.2

A snack box contains almonds, cashews, raisins and pistachios. The table below shows their weight and the percentage of their weight. What is the weight of almonds?

	Weight (in grams)	Weight Percentage (%)
Almonds		
Cashews	380	
Raisins	550	
Pistachios	320	20
TOTAL		100

Solution:

Let's find the weight percentage of raisins (r):

If   320 grams   20%
      550 grams   r%

$$r = \frac{550(20)}{320} = 34.375\%$$

Chart, Table and Data Analysis

Now, let's find the weight percentage of cashews (c):

If   320 grams   20%
          380 grams   c%

$$c = \frac{380(20)}{320} = 23.75\%$$

Our table looks like this now:

	Weight (in grams)	Weight Percentage (%)
Almonds		
Cashews	380	23.75
Raisins	550	34.375
Pistachios	320	20
TOTAL		100

We can easily find the weight percentage of almonds:

$$a = 100 - 20 - 34.375 - 23.75$$

$$a = 21.875\%$$

Finally, let's find the weight of almonds (m):

If   320 grams   20%
          m grams   21.875%

$$m = \frac{21.875(320)}{20}$$

$$m = 350 \; grams$$

Example 3.3

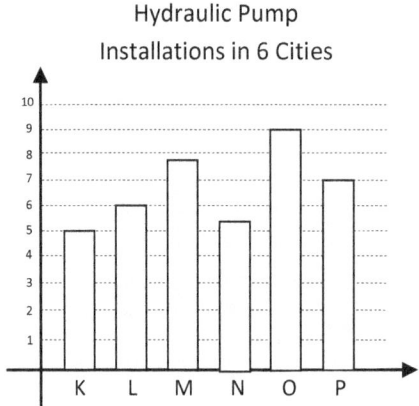

Hydraulic Pump
Installations in 6 Cities

The number of hydraulic pump installations in 6 cities is shown in the graph above. If the total number of installations is 401,000, then what is a suitable label for the vertical axis of the graph?

Solution:

The grids associated with our 6 cities are:

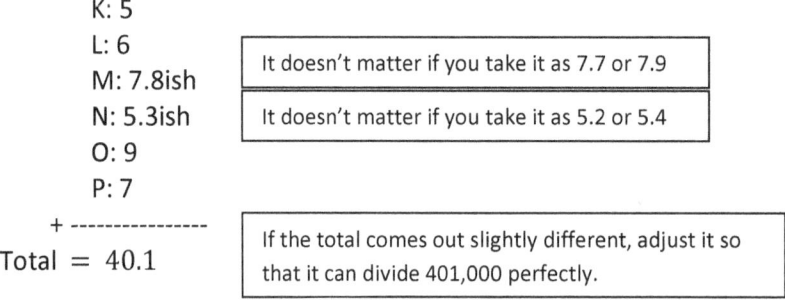

K: 5
L: 6
M: 7.8ish — It doesn't matter if you take it as 7.7 or 7.9
N: 5.3ish — It doesn't matter if you take it as 5.2 or 5.4
O: 9
P: 7
+ ----------------
Total = 40.1 — If the total comes out slightly different, adjust it so that it can divide 401,000 perfectly.

When we multiply the values on vertical axis by its label (y), we have the real value.

So, $40.1(y) = 401,000$

$y = \frac{401,000}{40.1}$

$y = 10,000$

Chart, Table and Data Analysis

Example 3.4

XYZ motorbike manufacturing factory produces X, Y and Z motorbike models every year. In 2017 the factory produced a total of 2400 motorbikes. The pie chart below shows the percentage of the number of motorbikes produced in 2017.

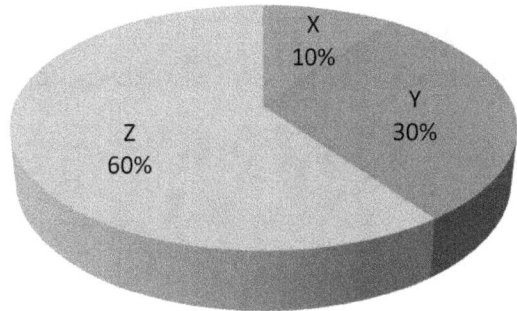

The bar chart below shows the ratio of the number of the motorbike models sold to the number of the same models produced in 2017. If the total number of motorbikes sold in 2017 was 1260, what is the ratio of the number of Z model motorbikes sold in 2017 to the number of Z model motorbikes produced in 2017?

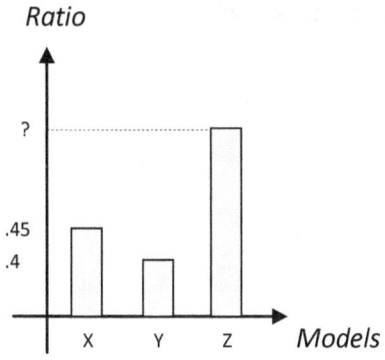

Solution:

Let's first find the number of motorbikes produced per model:

X : 10%, 2400(0.1) = 240

Y: 30%, 2400(0.3) = 720

Z: 60%, 2400(0.6) = 1440

We are done with the pie chart. Now, let's have a look at the bar chart, note down the ratios for each model and calculate how many motorbikes (for X and Y) were sold.

$$X: \frac{Number\ of\ X\ models\ sold}{Number\ of\ X\ models\ produced} = 0.45$$

$$X: \frac{Number\ of\ X\ models\ sold}{240} = 0.45$$

So, The number of X models sold was = 108

$$Y: \frac{Number\ of\ Y\ models\ sold}{Number\ of\ Y\ models\ produced} = 0.4$$

$$Y: \frac{Number\ of\ Y\ models\ sold}{720} = 0.4$$

So, The number of Y models sold was = 288

We know that the total number of motorbikes sold was 1260

Thus, The number of Z models sold was:

$$1260 - (108 + 288) = 864$$

Finally, $Z: \dfrac{Number\ of\ Z\ models\ sold}{Number\ of\ Z\ models\ produced} = ?$

$$Z: \frac{864}{1440} = \mathbf{0.6}\ \text{or}\ .6\ \text{or}\ \mathbf{60\%}$$

Chart, Table and Data Analysis

Example 3.5

A commercial kitchen uses meat (m), vegetables (v), and other ingredients (i) to cook 3 main dishes every day. First dish costs $400 to make and requires 20 kilograms of meat, 30 kilograms of vegetables and 40 kilograms of other ingredients. Second dish costs $230 to make and requires 10 kilograms of meat, 15 kilograms of vegetables and 30 kilograms of other ingredients. Third dish costs $225 to make and requires 15 kilograms of meat, 15 kilograms of vegetables and 20 kilograms of other ingredients. What is the cost of using 1 kilogram of vegetables?

1. dish	20m	+	30v	+	40i	=	$400
2. dish	10m	+	15v	+	30i	=	$230
3. dish	15m	+	15v	+	20i	=	$225

Solution:

$20m + 30v + 40i = 400$
$2 (10m + 15v + 30i = 230) 2$

$20m + 30v + 40i = 400$
$20m + 30v + 60i = 460$

If we can equate two variables to each other, only then can we isolate the

Let's pick 1. and 2. dishes

$-1 (20m + 30v + 40i = 400) -1$
$20m + 30v + 60i = 460$

$- 20m - 30v - 40i = - 400$
$20m + 30v + 60i = 460$

$20i = 60$
$i = 3$

$10m + 15v + 30i = 230$
$15m + 15v + 20i = 225$

Let's pick 2. and 3. dishes

$10m + 15v + 30(3) = 230$
$15m + 15v + 20(3) = 225$

We know i = 3

$-1\ (10m + 15v = 140)\ -1$
$15m + 15v = 165$

$5m = 25$
m = 5

$15m + 15v = 165$
$15(5) + 15v = 165$

> We can pick any equation we want.

$15v = 90$
v = 6   So, the cost of using 1 kilogram of vegetables is $6.

Practice Questions:

Questions 1-5 refer to the data below on revenue and expenditure statistics for Clucky's Clocks in year 2017.

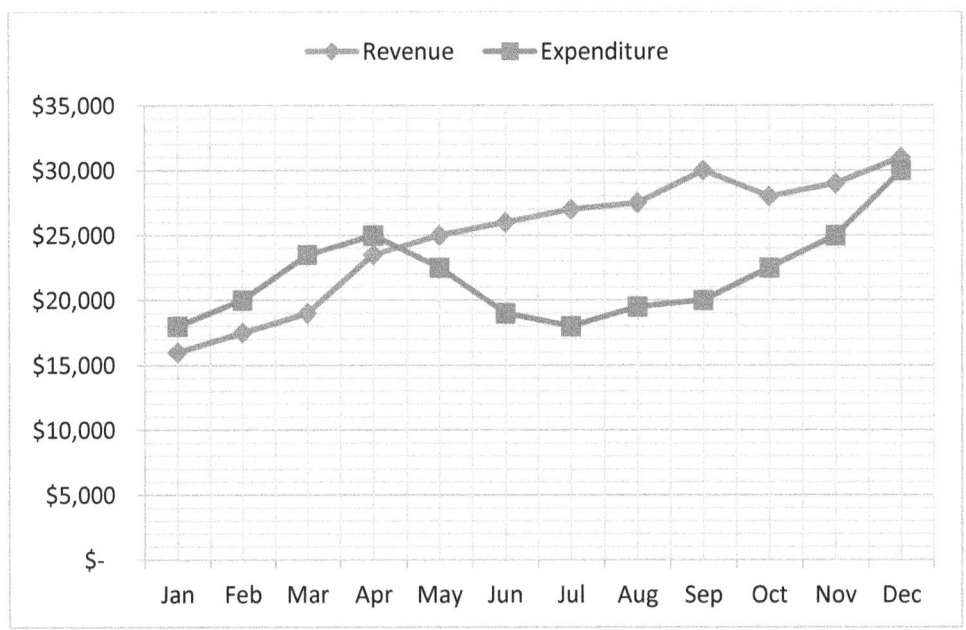

Chart, Table and Data Analysis

1) If the Months (i.e. Jan, Feb, ..., Dec) indicate the expenditure and revenue totals for each month by the end of the corresponding month, approximately when was Clucky's Clocks' revenue equal to its expenditures?
   a. Jan 15   b. Apr 14   c. May 7   d. Apr 24   e. May 14

2) Which of the following can be inferred from the graph? (Select ALL that apply)
   a. Since the beginning of the year, Clucky's Clocks' revenues have always risen gradually.
   b. Clucky's Clocks' revenue rose uniformly throughout the 12-month period
   c. Clucky's Clocks' revenue was never less than 75% of its expenses in 2017.
   d. The difference between Clucky's Clocks' revenue and expenditure has never been more than $10,000 in any of the months of the year 2017.
   e. The increase in Clucky's Clocks' revenue has a directly proportionate relationship with its decrease in expenditure after May 2017.

3) By what percent did Clucky's Clocks' expenditure increase from February to April?
   a. 40%   b. 20%   c. 25%   d. 10%   e. 45%

4) Between which two months was the percent increase in Clucky's Clocks' expenditure greatest?
   a. April and May
   b. November and December
   c. September and October
   d. January and February
   e. August and September

5) Which of the following is closest to Clucky's Clocks' average monthly expenditure for the third quarterly period (from July to September)?
   a. 16,174   b. 19,167   c. 15,548   d. 18,454   e. 20,547

6) If net income is defined as revenue minus expenditure, what is the best approximation to the difference in dollars between Clucky's Clocks' net income in April and September?
   a. 8,500   b. 10,000   c. 12,800   d. 11,500   e. 7,500

7) According to the chart below, if Audi produced 378,950 cars in Germany, how many of them would be sold in Germany?

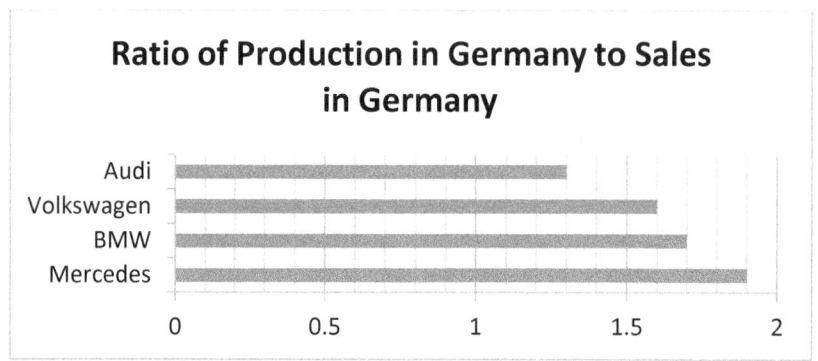

8) According to the same chart above, if BMW exported 243,880 cars that have been manufactured in Germany, what would be BMW's production in Germany? (Cars manufactured in Germany are either sold in Germany or exported to another country) (You may use a calculator)

9)

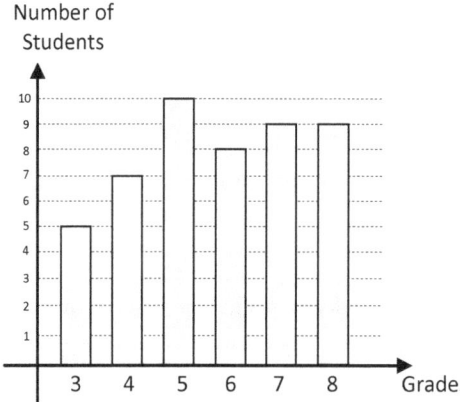

The number of students in a class and their grades are shown. If a student has to get at least 5 to pass, what is the percentage of the students who failed?

Chart, Table and Data Analysis

10) **Percent of citizens earning $60,000 or higher per annum**

Country	Percent of Citizens
Spain	36.48%
Germany	61.23%
France	54.85%
Poland	24.46%
Italy	42.74%
Greece	28.65%
Romania	14.29%
Bulgaria	12.41%

A survey was conducted among the citizens of the European Union member countries asking if they earn more than $60,000 or higher per annum. The results from 8 member countries are shown in the table above. The median percent of citizens who claimed to earn $60,000 or higher per annum for all 28 countries was 41.37%. What is the difference between the median percent of citizens who claimed to earn $60,000 or higher per annum for the 8 countries shown above and the median for all 28 countries?

11) Matthew collects, buys, sells and trades stamps. The graph below shows the number of stamps in his collection over 6 months. Between which two months did the number of stamps decline the fastest?

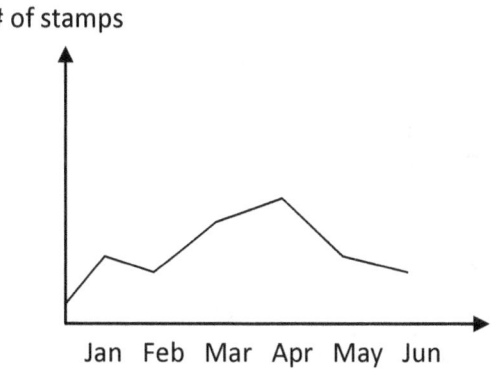

12)

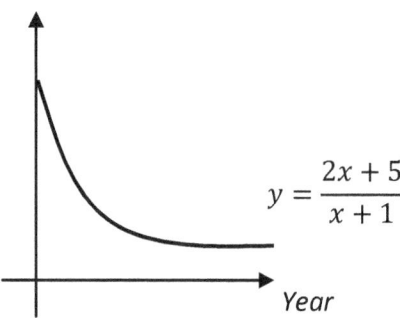

The graph above shows a bank's variable interest rates changing over years. After what year will the bank's interest rate fall under 3%?

    a)1   b)2   c)3   d)4   e)5   f)6

Which of the below hypothetically represents the highest and lowest interest rates that a bank customer can get from this bank?

    a){5, 1.99}   b){5, 2}   c){4, 3}   d){5, 2.99}   e){5, 2.01}

13)

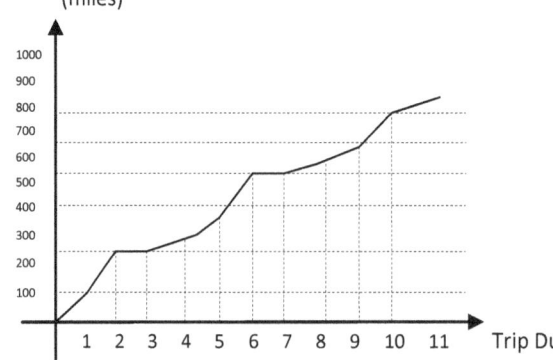

Chart, Table and Data Analysis 79

The graph above shows the duration of a bus trip and the distance traveled by the bus which took 2 breaks during the trip. How many hours are there between the beginning of the first break and the end of the second break?

14)

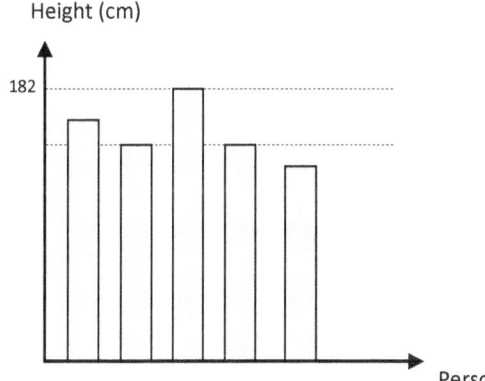

The graph above shows how tall 5 classmates are. If:
- Mary is as tall as Stephen;
- Barry is 2 cm shorter than Stephen;
- Elizabeth is 6 cm taller than Michael; and
- Michael is 3 cm taller than Mary,

What is the average height of these classmates?

15)

Profit/Loss (in thousands)

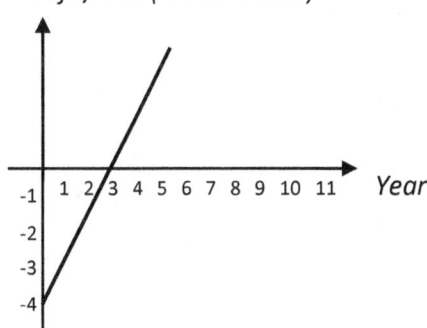

The graph above shows the Profit/Loss situation of a small company over 11 years. When there is no profit or loss, the company is said to break-even. What is the company's profit/loss situation 6 years after it breaks even?

a) $4000 profit   b) $8000 profit   c) $6000 profit   d) $2500 profit

16)

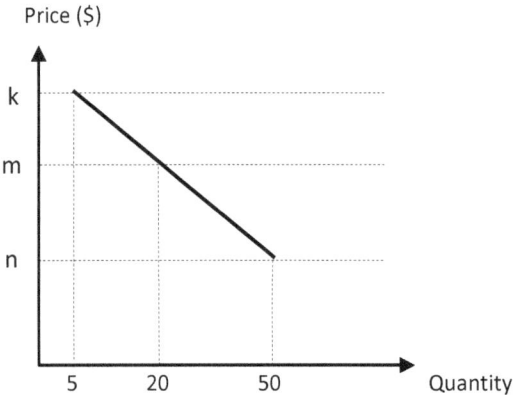

The demand line of a commodity is given in the graph above. What is the value of $(k - m)$?

17)

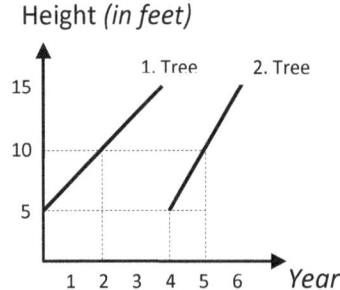

The graph shows the growth of 2 trees planted 4 years apart. How many years after the plantation of the second tree will it take for both trees to have equal heights?

CHAPTER IV

Functions

Example 4.1

If $(x) = \frac{2x^2-3x+1}{1-x}$, what is $f(-1)$?

Solution:

Let's just replace -1 with x

$$f(-1) = \frac{2(-1)^2-3(-1)+1}{1-(-1)}$$

$$f(-1) = \frac{2+3+1}{1+1} = \frac{6}{2}$$

$$f(-1) = 3$$

Example 4.2

If $(x) = |x-1| - |x| + 1$, what is the value of $f(1) + f(0) + f(-1)$?

Solution:

$f(1) = |1-1| - |1| + 1$

$f(1) = 0$

$f(0) = |0-1| - |0| + 1$

$f(0) = 2$

> The absolute value $|k|$ of a real number k is the non-negative value of k regardless of its sign.

$$f(-1) = |-1-1| - |-1| + 1$$

$$f(-1) = 2 - 1 + 1$$

$$f(-1) = 2$$

$$f(1) + f(0) + f(-1) = 0 + 2 + 2 = 4$$

Example 4.3

If (a, b) satisfies the system of equations below, and $a \neq 0$; and $b \neq 0$, what is the value of $a + b$?

$$\left|\frac{a}{b}\right| = 5a$$

$$|ab| = -3b$$

Solution:

First, let's consider the impact of the absolute value signs:

a must be positive ($a > 0$), because $\left|\frac{a}{b}\right| = 5a$

b must be negative ($b < 0$), because $|ab| = -3b$

Now, we can ignore the absolute value signs.

$$\frac{a}{b} = 5a$$

$$ab = -3b$$

----

a = 5ab

$b = -\frac{1}{5}$ (b must be negative)

Functions

$$ab = -3b$$

$$a = 3 \quad \text{(a must be positive)}$$

$$a + b = 3 - \frac{1}{5} = \frac{14}{5}$$

Example 4.4

If (a, b, c) satisfies the equations below, what is the value of $a + b + c$?

$$|a| = 3 \ , \ |b| = 6 \ , \ |c| = 8$$

$$c < a < b$$

$$a.b.c > 0$$

Solution:

> a is either 3 or -3
>
> b is either 6 or -6
>
> c is either 8 or -8

In order for c to be less than both *a* and *b*, c must be negative.

So; $c = -8$

Since: $a.b.c > 0$; and $c = -8$; $(a.b)$ *must be negative*.

So; $a = -3$, and $b = 6$ | Consider: $c < a < b$ |

Finally $a + b + c = -3 + 6 - 8$

$$\boldsymbol{a + b + c = -5}$$

Example 4.5

The functions f and g are defined below. Show the expression that is equivalent to $\dfrac{f(x)}{g(x)}$

$$f(x) = 2x^2 - 7x - 15$$

$$g(x) = x^2 - 9$$

Solution:

$$\frac{f(x)}{g(x)} = \frac{2x^2-7x-15}{4x^2-9}$$

$$\frac{f(x)}{g(x)} = \frac{(2x+3)(x-5)}{(2x-3)(2x+3)}$$

$$\frac{f(x)}{g(x)} = \frac{x-5}{2x-3}$$

Example 4.6

Find the value of $f(4).g(4)$?

$$(f+g)(x) = x^2$$

$$(f-g)(x) = x$$

Solution:

$$(f+g)(x) = f(x) + g(x) = x^2$$

$$(f-g)(x) = f(x) - g(x) = x$$

$$f(4) + g(4) = 4^2 = 16$$

$$f(4) - g(4) = 4$$

Functions

$$2f(4) = 20$$

$$f(4) = 10$$

$$f(4) - g(4) = 4$$

$$g(4) = 6$$

$$f(4).g(4) = 10.6 = \mathbf{60}$$

Example 4.7

Find the value of $f(4).g(4)$?

$$(f+g)(x) = x^2$$

$$(f-g)(2x) = x$$

Solution:

$$(f+g)(x) = f(x) + g(x) = x^2$$

$$(f-g)(2x) = f(2x) - g(2x) = x$$

$$f(4) + g(4) = 16$$

$$f(4) - g(4) = 2$$

Warning! Make sure for both equations we are dealing with $f(4)$ values. So, for the first equation x must be 4, for the second x must be 2.

+ ---

$$2f(4) = 18$$

$$f(4) = 9$$

$$f(4) - g(4) = 2$$

$$g(4) = 7$$

$$f(4).g(4) = 9.7 = \mathbf{63}$$

Example 4.8

The function f is defined below. Find the value of $f^{-1}(2)$.

$f(x) = 3x - 1$

Solution:

$f^{-1}(x)$ is the inverse function of $f(x)$

> **Watch out!**
> $f^{-1}(x) \neq f(x)^{-1}$

$f(x) = y = 3x - 1$

Let's reverse x and y:

$x = 3y - 1$

Now, leave y alone:

$y = \frac{x+1}{3} = f^{-1}(x)$

> When we reverse x and y;
> our new $y = f^{-1}(x)$

$f^{-1}(x) = \frac{x+1}{3}$

$f^{-1}(2) = \frac{2+1}{3} = 1$

Example 4.9

The functions f and g are defined below. What is the value of $g(1)$?

$$f(x) = 3x + 2$$

$$g(x) = x - f(x)$$

Solution:

$g(x) = x - f(x)$ — Let's replace 1 with x inside the g function.

$g(1) = 1 - f(1)$ — We need to find $f(1)$.

Functions

$f(x) = 3x + 2$

$f(1) = 3 + 2$

$f(1) = 5$

> Now we replace 1 with x inside the f function.

$g(1) = 1 - f(1)$

$g(1) = 1 - 5$

$\boldsymbol{g(1) = -4}$

Example 4.10

The functions f and g are defined below. What is the value of $g(5)$?

$f(x) = x^2 + 2$

$g(x) = f(x) - 2$

Solution:

$g(5) = f(5) - 2$

$f(5) = 5^2 + 2$

$f(5) = 27$

$g(5) = 27 - 2$

$\boldsymbol{g(5) = 25}$

Example 4.11

The functions f and g are defined below. What is the value of $f(1)$?

$f(g(x)) = 2 - g(x)$

Solution: $g(x) = 1$

$f(1) = 2 - 1$

$f(1) = 1$

> In order to find $f(1)$ we need to give the value of 1 to $g(x)$.

Example 4.12

Find $f(3)$, when $f(x - 1) = x^2 - 2x + 5$

Solution:

To find $f(3)$, the value inside the parenthesis should be 3.

$x - 1 = 3$

So, $x = 4$

$f(4 - 1) = 4^2 - 2(4) + 5$

> Let's replace x with 4

$f(3) = 16 - 8 + 5$

$f(3) = 13$

Example 4.13

Find $g(7)$, when $g(x^2 - 2) = 3x + 10$, and $x < 0$

Solution:

To find $g(7)$, the value inside the parenthesis should be 7.

$x^2 - 2 = 7$

$x^2 = 9$

$x = \pm 3$

Functions 89

So, $x = -3$ [We know that $x < 0$]

$g((-3)^2 - 2) = 3(-3) + 10$ [Let's replace x with -3]

$g(7) = 1$

Example 4.14

If $x \neq 2$, find $f(1)$, when:

$$f(x) = \frac{x^2 + 3x - 1}{x^3 - 2x}$$

Solution:

$$f(1) = \frac{1^2 + 3(1) - 1}{1^3 - 2(1)}$$

$$f(1) = \frac{1 + 3 - 1}{1 - 2}$$

$$f(1) = \frac{3}{-1} = -3$$

Example 4.15

If $g(x) > 0$, find $f(3)$,

when: $\begin{cases} f(x) = kx^2 + 2 \\ g(x) = \sqrt{x} + \frac{x}{8} \\ f(g(16)) = 8 \end{cases}$

Solution:

Let's find $g(16)$:

$g(16) = \sqrt{16} + \dfrac{16}{8}$ $\quad$ | $\sqrt{16} = \pm 4$, however since we know that $g(x) > 0$
$g(16) = 4 + 2 = 6$ $\quad$ | $\sqrt{16} = 4$

$f(g(16)) = 8$

$f(6) = 8$

$f(x) = kx^2 + 2$

$f(6) = k6^2 + 2$

$8 = k6^2 + 2$

$6 = k6^2$

$k = \dfrac{1}{6}$

So, $\quad f(x) = kx^2 + 2$

$\quad f(x) = \dfrac{x^2}{6} + 2$

$\quad f(3) = \dfrac{3^2}{6} + 2 = \dfrac{9}{6} + 2 = \dfrac{3}{2} + 2$

$\quad f(3) = \dfrac{7}{2}$

Example 4.16

Find $\dfrac{f(h(0))}{h(f(1))}$, when $f(x) = \dfrac{1}{x^2+1}$ and $h(x) = \sqrt{x+2}$

Solution:

Let's find $h(0)$ first and then $f(1)$

$h(0) = \sqrt{0+2}$ $\qquad$ $f(1) = \frac{1}{1^2+1}$

$h(0) = \sqrt{2}$ $\qquad$ $f(1) = \frac{1}{2}$

$\dfrac{f(h(0))}{h(f(1))} = \dfrac{f(\sqrt{2})}{h\left(\frac{1}{2}\right)}$

Let's find $f(\sqrt{2})$ and $h\left(\frac{1}{2}\right)$

$f(\sqrt{2}) = \dfrac{1}{\sqrt{2}^2+1}$ $\qquad$ $h\left(\dfrac{1}{2}\right) = \sqrt{\dfrac{1}{2}+2}$

$f(\sqrt{2}) = \dfrac{1}{3}$ $\qquad$ $h\left(\dfrac{1}{2}\right) = \sqrt{\dfrac{5}{2}}$

$\dfrac{f(\sqrt{2})}{h\left(\frac{1}{2}\right)} = \dfrac{\frac{1}{3}}{\sqrt{\frac{5}{2}}} = \dfrac{1}{3} \cdot \dfrac{\sqrt{2}}{\sqrt{5}} = \dfrac{\sqrt{2}}{3\sqrt{5}}$

$\dfrac{\sqrt{2}}{3\sqrt{5}} = \dfrac{\sqrt{10}}{15}$

$(\sqrt{5})$

Example 4.17

The number of bacteria in a small sample of culture after k minutes is given by:

$$N = f(k) = 900 \left(\dfrac{5}{3}\right)^k$$

What is the percentage of bacterial growth between the first and second minute?

Solution:

$$N_1 = f(1) = 900\left(\frac{5}{3}\right)^1 = 1500$$

$$N_2 = f(2) = 900\left(\frac{5}{3}\right)^2 = 2500$$

So, the bacterial growth in numbers is $N_2 - N_1$

Bac. Growth: $2500 - 1500 = 1000$

The percentage of growth/increase must be based on the initial amount (N_1).

Bac. Growth (%) : $1500 \cdot \frac{x}{100} = 1000$

$$x = \frac{1000}{15}$$

$$x = \%66.7$$

Example 4.18

A factory determines that the total number of products being manufactured, P, is a function of the number of employees, n, where:

$$P = f(n) = \frac{50n - n^2}{7}$$

The total revenue, R, made by selling k units is given by the function, h, where:

$$R = h(k) = 56k$$

a) Find $(h \circ f)(n)$ or $h(f(n))$

b) If there were 40 employees, and all of the manufactured items were sold, what would be the revenue?

Functions

Solution:

a) $(h \circ f)(n)$ or $h(f(n))$

$h(f(n)) = h\left(\dfrac{50n - n^2}{7}\right)$ — Plug $f(n)$ into the function of h.

$h(k) = 56k$

$h(f(n)) = 56\left(\dfrac{50n - n^2}{7}\right)$ — Replace k with $f(n)$

$h(f(n)) = 8\left(\dfrac{50n - n^2}{1}\right)$

$h(f(n)) = 400n - 8n^2$

b) $R = h(k) = h(f(n)) = 400n - 8n^2$
$R = 400 \cdot 40 - 8 \cdot 40^2$
$R = 3200$

Practice Questions

1) If (a, b) satisfies the system of equations below, what is the value of $a + b$?
$b - a = 2$
$b - |a - b| = 1$

2) Find $g(f(-2))$, when $f(x) = x^2$ and $g(x) = 3x - 2$

3) Find $g^{-1}(f(-2))$, when $f(x) = x^2$ and $g(x) = 3x - 2$

4) Find $f(1)$, when $f(x + 2) = x^3 - 3x^2 - x + 1$

5) Find the value of g(1).
$$f(g(x)) = f(x) \cdot g(x)$$
$$f(x) = 3x + 4$$

6) Find $\frac{f(g(3))}{g(f(1))}$, when $f(x) = \frac{2x^2 - 3}{x}$ and $g(x) = \frac{5x - f(x)}{2}$

7) The population of a small village grows at the rate, r, of 7.2% per year since 1995. The population is a function of t which can be formulized as follows:

$$P = f(t) = IP(1 + r)^t \quad , \text{where:}$$
P: number of new population
IP: number of initial population
r: population growth rate per year
t: number of years passed since the initial year

If the population was 12,502 in 1999, what would be the population in 2017?

8) An online business with an existing capital of X has revenue and expenses each month of Y and Z, respectively. If all profits are kept in the firm, show the liquid value, LV, of the firm at the end of n months as a function of n. ($Profit = Revenue - Expenses$)

9) A study has demonstrated statistical relationship between an individual's status, income and education. Status, S, has been determined to be a function of income, I, and income to be a function of years of education, E.

$$S = f(I) = 0.42(I - 1000)^{0.76}$$
$$I = g(E) = 5216 + 0.21E^{2.45}$$

Functions

What is the status difference between an individual who received 15 years of education and another individual who received 10 years of education?

10) Find $f(3)$, when $f\left(\frac{x-2}{x+2}\right) = x^2 - 2x + 2$

11) Find $g(1)$, when $(f \circ g)(x) = f(x) \cdot g(x)$ and $f(x) = 2x + 3$

12) Find $f(3)$, when:
$f(x) = ax^2 + 1$
$g(x) = \sqrt{x} + 2$
$(f \circ g)(9) = 6$

13) Find $(f \circ g^{-1})(x)$, when $g(x) = 3x - 6$ and $f(x) = (x - 2)^2$

14) For f function, k is a positive integer; $k \geq 1$. If:
$f(k) = 2f(k - 1) + 1$;
$f(0) = 1$, what is the value of $f(3)$?

15) If $f(x) = 3^{x+2}$, which of the following expressions is equal to $f(a + b - 1)$?

a) $\frac{f(a+b)}{27}$ b) $\frac{f(a+b)}{9}$ c) $\frac{f(a).f(b)}{9}$ d) $\frac{f(a).f(b)}{27}$ e) $\frac{f(a).f(b)}{81}$

16) What is the sum of all possible values of x that can satisfy the function of equations below?
$f(x) = |3x - 1|$
$g(x) = |x + 1|$
$(g \circ f)(x) = 4$

17) The graph of $f(x)$ is shown below. Point $A(-1, 6)$ and Point $B(1, -4)$ are on $f(x)$ parabola which crosses x-axis at 2, 0 and -3.

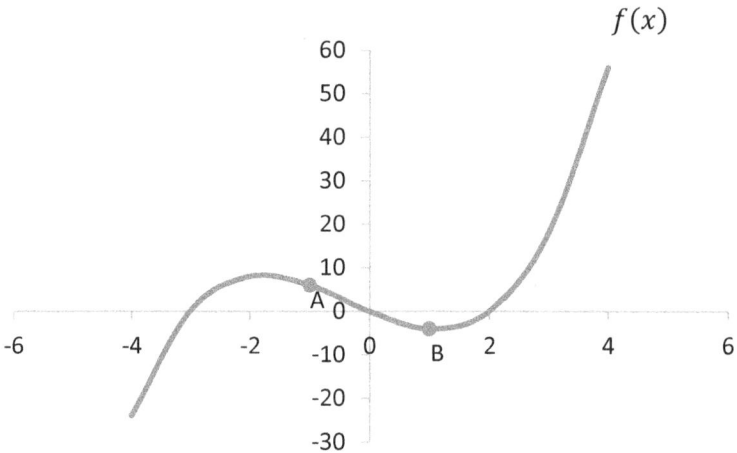

If $k(x) = x - f(x-1)$, what is the value of $k(3) - k(0)$?

18) The graph of $f(x)$ is shown below. Point $A(2, -4)$ is on symmetrical $f(x)$ parabola which crosses x-axis at 4, 0 and -2.

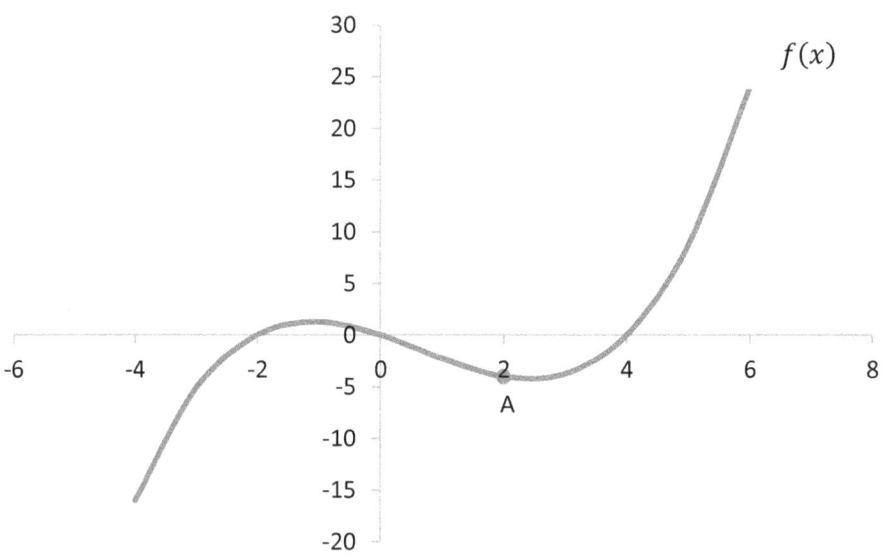

If $g(x) = f(2x - 1) + 2$, what is the value of $g(2)$?

CHAPTER V

Inequalities, Statistics and Complex Numbers

Example 5.1 **Inequalities**

If x and y are both positive integers that satisfy the inequality below, what is the sum of the values that y can have?

$$1 < x < y - x < 5$$

Solution:

By looking at the inequality, we can say that x must be either 2 or 3. So;

If $x = 2$ $\begin{cases} 1 < 2 < 3 < 5 : y - x = 3; y = 5 \\ 1 < 2 < 4 < 5 : y - x = 4; y = 6 \end{cases}$

If $x = 3$ $\{1 < 3 < 4 < 5 : y - x = 4; y = 7$

Let's add up all the possible values of y.

$$5 + 6 + 7 = 18$$

Example 5.2

If x is an integer which satisfies the system of equation below, what is the product of the maximum and minimum values of $(2 - x)$?

$$-3 < x < 5$$

Solution:

We need to turn the above equation into the format of $2 - x$

Let's first turn everything into negative:

$-(-3 < x < 5)-$ When everything gets turned into negative, inequality signs must be reversed.

$3 > -x > -5$

Let's add 2 up, so as to make it $2 - x$

$+2(3 > -x > -5) + 2$

$5 > 2 - x > -3$

$Max(2 - x) = 4 \; ;$

$Min(2 - x) = -2$

$Max . Min = -8$

Example 5.3

What is the sum of the integers of x that satisfy the system of inequality below?

$|x + 3| \leq 5$

Solution:

We need to get rid of the absolute value ($|...|$)

$|x + 3|$ *is either positive or negative* $|x + 3| = \pm(x + 3)$

We need to consider both possibilities.

When $|x + 3|$ *is positive;* $|x + 3| = x + 3$

So; $x + 3 \leq 5$

When $|x + 3|$ *is negative;* $|x + 3| = -x - 3$

So; $-x - 3 \leq 5$ or $x + 3 \geq -5$ or $-5 \leq x + 3$

Inequalities, Statistics and Complex Numbers

Let's combine our two possibilities into one:

$$-5 \leq x + 3 \leq 5$$

Now; let's get rid of "+3"

$$-5 - 3 \leq x + 3 - 3 \leq 5 - 3$$

$$-8 \leq x \leq 2$$

So; $\Sigma x = \{-8 - 7 - 6 - 5 - 4 - 3 - 2 - 1 + 0 + 1 + 2\}$

$\Sigma x = -33$

Example 5.4

What is the solution interval of the values of x that satisfy the system of inequality below?

$$x^2 - 3x + 2 < 0$$

Solution:

Let's factorize the inequality as if it is a normal equation.

$$x^2 - 3x + 2 = 0$$
$$\begin{array}{ll} x & -2 \\ x & -1 \end{array}$$

$$(x - 2)(x - 1) = 0$$

We find two roots: 2 and 1. Let's place them in our xy-table.

x	$-\infty$	1	2	$+\infty$
y				

We need to assign different values to x so as to see whether y turns out positive or negative.

Let's assign 5 to x:

$$(5-2)(5-1) = 12$$

Since $12 > 0$; we can say that for all values of x between 2 and $+\infty$, the value of y will be positive.

x	$-\infty$		1		2		$+\infty$
y						+	

Let's assign 0 to x:

$$(0-2)(0-1) = 2$$

Warning! We can assign any value we want here so long as it is less than 1.

Since $2 > 0$; we can say that for all values of x between $-\infty$ and 1, the value of y will be positive.

x	$-\infty$		1		2		$+\infty$
y		+				+	

Let's assign 1.5 to x:

$$(1.5-2)(1.5-1) = -0.25$$

Since $-0.25 < 0$; we can say that for all values of x between 1 and 2, the value of y will be negative.

x	$-\infty$		1		2		$+\infty$
y		+		−		+	

$$x^2 - 3x + 2 < 0$$

We are looking for the negative values of y.

So, $1 < x < 2$

Warning! 1 and 2 are not inclusive.

Example 5.5

What is the solution interval of the values of x that satisfy the system of inequality below?

$$(x+1)^2 < 4$$

Solution:

Let's factorize the inequality as if it is a normal equation.

$(x+1)^2 - 4 = 0$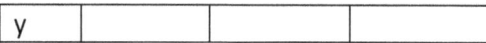

$(x+1-2)(x+1+2) = 0$

$(x-1)(x+3) = 0$

$x_1 = 1 \; ; \; x_2 = -3$

x	$-\infty$	-3	1	$+\infty$
y				

$(x+1)^2 - 4 < 0$

Let's assign values to x:

$x = 2 \Rightarrow y = 5 > 0$

$x = -1 \Rightarrow y = -4 < 0$

$x = -6 \Rightarrow y = 21 > 0$

x	$-\infty$	-3	1	$+\infty$
y	$+$	$-$	$+$	

$(x+1)^2 - 4 < 0$

We are looking for the negative values of y.

So, $-3 < x < 1$ | **Warning!** 1 and -3 are not inclusive. |

Example 5.6   Statistical Analysis

The participants of a psych-test were divided into two groups: Test Group and Control Group. There were 20 participants in the Test Group who scored an average psych-test result of 35 points and there were 30 participants in the Control Group who scored an average psych-test result of 40 points. What was the average score of all participants who took the psych-test?

Solution:

Test Group: Average result $= \dfrac{Test\ Group's\ Total\ results}{20} = 35$

Control Group: Average result $= \dfrac{Control\ Group's\ Total\ results}{30} = 40$

Average Score of All Participants
$= \dfrac{Test\ Group's\ Total\ results\ +\ Control\ Group's\ Total\ results}{20\ +\ 30}$

Average Score of All Participants $= \dfrac{35 \cdot 20\ +\ 40 \cdot 30}{20\ +\ 30} = \dfrac{20(35+60)}{50} = \dfrac{2(95)}{5}$

Average Score of All Participants = 38

Example 5.7

The arithmetic mean of the results Thomas scored at 2 tests was 6. If his final grade will be the arithmetic mean of the 3 tests, what should he at least score in the third test to guarantee the final grade of 7?

Solution:

Arithmetic Mean (Average) of the first 2 Tests $= \dfrac{2\ Tests'\ Total\ results}{2} = 6$

Inequalities, Statistics and Complex Numbers

So, 2 Tests' Total Results $= 2 \cdot 6 = 12$

Arithmetic Mean (Average) of the 3 Tests $= \frac{3\ Tests'\ Total\ results}{3} \geq 7$

So, 3 Tests' Total Results $\geq 3 \cdot 7$

2 Tests' Total Results + Third Test's Result ≥ 21

So, $12 +$ Third Test's Result ≥ 21

Third Test's Result ≥ 9

Example 5.8

In a warehouse there were 9 gray containers each weighing 3 tonnes and 12 purple containers each weighing 6 tonnes. Some of the gray and purple containers were moved by staff to another warehouse which was empty.

After this new arrangement the average weight of the containers in the first warehouse was 5 tonnes and the average weight of the containers in the second warehouse was 4 tonnes. How many purple containers were moved from the first warehouse to the second?

Solution:

Let's say we moved x number of gray containers and y number of purple containers.

	Before 1st W.house		After 1st W.house		Before 2nd W.house		After 2nd W.house	
	#	Weight	#	Weight	#	Weight	#	Weight
Gray	9	3(9)	$9-x$	$3(9-x)$	0	0	x	$3x$
Purple	12	6(12)	$12-y$	$6(12-y)$	0	0	y	$6y$

Average weight = $\frac{Total\ Weight}{Total\ \#\ of\ containers}$

After the arrangement, for the 1ˢᵗ Warehouse:

Average weight = $\frac{27-3x+72-6y}{9-x+12-y} = 5$

$\frac{99-3x-6y}{21-x-y} = 5$

$99 - 3x - 6y = 105 - 5x - 5y$

$2x - y = 6$

$\boldsymbol{y = 2x - 6}$

After the arrangement, for the 2ⁿᵈ Warehouse:

Average weight = $\frac{3x+6y}{x+y} = 4$

$4x + 4y = 3x + 6y$

$\boldsymbol{x = 2y}$

So; $x = 2y$

   $y = 2x - 6$

> Let's replace **x** with **2y**, since

   $y = 2(2y) - 6$

   $y = 4y - 6$

   $3y = 6$

   $\boldsymbol{y = 2}$

> **Remember!**
> **y** represents the number of purple containers that we moved from the 1ˢᵗ to the 2ⁿᵈ warehouse.

Inequalities, Statistics and Complex Numbers

Example 5.9   **Complex Numbers**

$i = \sqrt{-1}$, Find: $(11 - 4i) - (16i^2 - 5i)$

Solution:

Warning!
$16i^2 \neq (16i)^2$

Since, $i = \sqrt{-1} \leftrightarrow i^2 = -1$

$(11 - 4i) - (16(-1) - 5i) =$

$(11 - 4i) - (-16 - 5i) =$

$11 - 4i + 16 + 5i =$

$\mathbf{27 + i}$

Example 5.10

$i = \sqrt{-1}$, what is the ratio of $1 + i$ to $1 - i$?

Solution:

$$\frac{1+i}{1-i} = ?$$

Let's expand both numerator and denominator by $(1 + i)$

$$\frac{1+i}{1-i} = \frac{(1+i)(1+i)}{(1-i)(1+i)} = \frac{1+2i+i^2}{1-i^2}$$

$(1 + i)$

Since, $i = \sqrt{-1} \leftrightarrow i^2 = -1$

$$\frac{1+2i+i^2}{1-i^2} = \frac{1+2i-1}{1--1} = \frac{2i}{2} = i$$

Example 5.11

Find the equation with roots 3, $2-i$ and $2+i$.

Solution:

$$(x-3)[x-(2-i)][x-(2+i)] = 0$$

$$(x-3)(x-2+i)(x-2-i) = 0$$

$$(x-3)(x^2 - 2x - xi - 2x + 4 + 2i + xi - 2i - i^2)$$

$$(x-3)(x^2 - 4x + 5)$$

$$x^3 - 4x^2 + 5x - 3x^2 + 12x - 15$$

$$x^3 - 7x^2 + 17x - 15$$

Practice Questions:

1. If k is an integer which satisfies the system of equation below, what is the product of the maximum and minimum values of $(k+1)$?
$$-7 < 2k < 8$$

2. What is the solution interval of the values of x that satisfy the system of inequality below?
$$x^2 < 2x + 3$$

3. What is the solution interval of the values of x that satisfy the system of inequality below?
$$(5-x)(3x-1) > 0$$

4. $-3 < 2 - x < 3$; $-2 \leq y - 1 \leq 4$
If (x, y) integers satisfy the system of inequality above, what is the least possible value of: $|2x - 3| + y$?

Inequalities, Statistics and Complex Numbers

5. What is the solution interval of the values of x that satisfy the system of inequality below?
$$\frac{x(x^2+4x+4)}{3-x} \geq 0$$

6. What is the solution interval of the values of x that satisfy the system of inequality below?
$$\frac{(x+3)(2-x)}{x} > 0$$

7. What is the smallest possible integer value of x that satisfies the inequality below?
$$\frac{-(x+5)(x+6)^2}{x} > 0$$

8. What is the sum of the integer values of x that satisfy the inequality below?
$$(7-x)(x+6)(x-5)^2 > 0$$

9. What is the sum of the integer values of x that satisfy the inequality below?
$$\frac{x^2-8x+7}{(x+2)^2} < 0$$

10. What is the sum of the integers of x that satisfy the system of inequality below?
$$|x-2| \leq 3$$

11. What is the smallest possible integer value of m that satisfies the inequality below?
$$\frac{m}{2} - \frac{3}{m} > 0$$

12. $i = \sqrt{-1}$, if x satisfies the system of equation below, what is the value of x in terms of i?

$$\frac{1+2i}{1-2i} = \frac{x+3}{5i^2}$$

13. The average profit made by a company in 2014, 2015 and 2016 was $5million. The company made 25% more profit in 2017 than 2016. If the average profit made in these 4 years was $6million, how much profit did the company make in 2016?

14. In a graduate school, the average age of students studying management, economics and accounting majors is 20, 26 and 29; respectively. The average age of management and economics students together is 23; and the average age of economics and accounting students is 28. What is the average age of all students studying all of the three majors?

15. The number of students taking Professor Matt Matix's class and their letter grades is given in Table 1 and the conversion of Points to Letter Grades is given in Table 2 below. What is the average letter grade of this class?

Table 1

Letter Grade	A	B	C	D	F	NG
Number of Students	12	10	5	2	1	0

Table 2

Letter Grade	A	B	C	D	F	NG
Point	5	4	3	2	1	0

CHAPTER VI

Coordinate Geometry and Basic Shapes

Example 6.1   Coordinate Geometry

In the xy- plane, the point (h, k) lies on the line with the equation $y = 2x + b$, where b is a constant. The point with coordinates ($3h$, $2k$) lies on another line with the equation $y = 3x + b$. If $k \neq 0$, what is the value of $\dfrac{h}{k}$?

Solution:

For the first line: (h, k) = (x, y)

$$y = 2x + b$$

$$k = 2h + b$$

$$\boldsymbol{b = k - 2h}$$

For the second line: (3h, 2k) = (x, y)

$$y = 3x + b$$
$$2k = 3(3h) + b$$
$$2k = 9h + b$$
$$\boldsymbol{b = 2k - 9h}$$

Let's isolate b, so we can equate 2 lines to each other.

$$b = k - 2h = 2k - 9h$$
$$k - 2h = 2k - 9h$$
$$7h = k$$

$$\frac{h}{k} = \frac{1}{7}$$

Example 6.2 Slope of a Line

m is the slope of this two variable (x, y) linear equation;

$$y = mx + b$$

Find the slope of the equations below?

$y = 2x + 1$ (Slope is 2)

$y = 3x - 2$ (Slope is 3)

$y = -2x + 14$ (Slope is -2)

$k = -4h - 5$ (Slope is -4)

$t = p - 3$ (Slope is 1)

$f(x) = x + 3$ (Slope is 1)

$f(t) = 2t - 5$ (Slope is 2)

$y = 4 - 2x$ (Slope is -2)

Example 6.3 Equation of Lines: Knowing one Point and the Slope

Find the equation of a line that has a slope of 3 and passes through (2, -1).

Solution:

$$m = \frac{y - y_1}{x - x_1}$$

(2, -1) is the point which corresponds to (x_1, y_1).

$$m = \frac{y - y_1}{x - x_1}$$

$$3 = \frac{y - (-1)}{x - 2}$$

$$3 = \frac{y+1}{x-2}$$

$$3x - 6 = y + 1$$

$$y = 3x - 7$$

Example 6.4   Equation of Lines: Knowing two Points

(x_1, y_1) and (x_2, y_2) are two different points on a line. What is the slope of the line?

Let us define m as the slope of the line.

$$m = \frac{y_2 - y_1}{x_2 - x_1} \quad \text{or} \quad m = \frac{y_1 - y_2}{x_1 - x_2}$$

Example 6.5

Find the equation of a line that passes through (2, -1) and (3, 1).

Solution:

$y = mx + b$, (where m is the slope and b is the constant)

When 2 points of a line are known:

$$m = \frac{y_2 - y_1}{x_2 - x_1}$$

$x_1 : 2$
$y_1 : -1$
$x_2 : 3$
$y_2 : 1$

$$m = \frac{1 - (-1)}{3 - 2}$$

$$m = 2$$

Let us plug "2" in its place: $y = 2x + b$

We still have to find "b". Let us run the coordinates of one of our points.

(2, -1)	or	(3, 1)
$-1 = 2(2) + b$		$1 = 2(3) + b$
$-1 = 4 + b$		$1 = 6 + b$
$b = -5$		$-5 = b$

So, the equation of our line is: $y = 2x - 5$

Example 6.6

Find the equation of the line below:

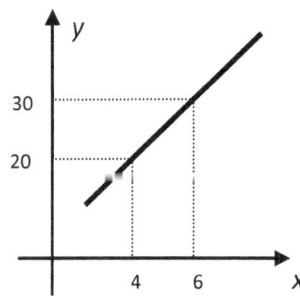

Solution:

The Points of the line are: (4, 20) and (6, 30)

$$m = \frac{y_2 - y_1}{x_2 - x_1}$$

$$m = \frac{30 - 20}{6 - 4} = 5$$

$y = mx + b$

$y = 5x + b$

Let's run the coordinates of one of our points, such as (4, 20)

$20 = 5(4) + b$

$b = 0$

So, the equation of the line is:

$y = 5x$

Example 6.7

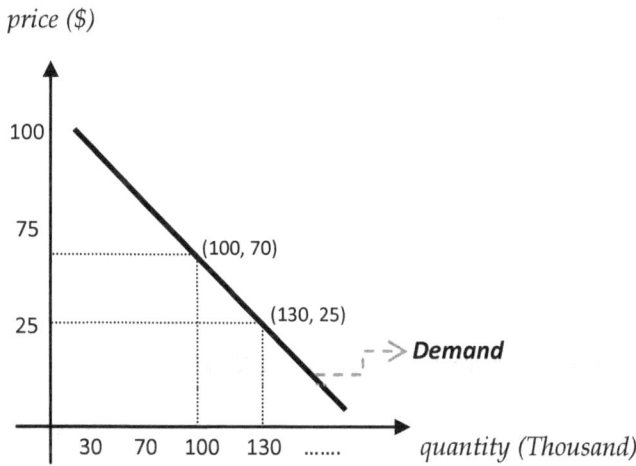

The demand line above demonstrates the relationship between the price (p) of a lamp (in $) and the quantity (q) of lamps (in thousands) that customers are willing to buy at that corresponding price. It is known that 100,000 lamps were bought at $70 each and 130,000 lamps were bought at $25 each.

Question 1: Which of the below statements is TRUE?

    a) 1 unit increase in q results in 1 unit increase in p.
    b) 1 unit decrease in q results in 1 unit increase in p.
    c) 1 unit increase in q results in 1.5 units increase in p.
    d) 1 unit increase in q results in 1.5 units decrease in p.

Question 2: At what price would 140,000 lamps be bought by customers?

Solution:
If we find the slope and the equation of the line, we can easily solve both questions.

The Points of the line are: (100, 70) and (130, 25)

$$m = \frac{y_2 - y_1}{x_2 - x_1}$$

$$m = \frac{25 - 70}{130 - 100} = \frac{-45}{30} = \frac{-3}{2} = -1.5$$

Now, we can answer the first question. Since the slope is negative, we can say that there is a negative correlation between q and p. When our base variable (q) increases by 1 unit, the second variable (p) will decrease by 1.5 units. So, the correct answer is D.

$$y = mx + b$$

$$y = -1.5x + b$$

Let's run the coordinates of one of our points, such as (100, 70)

$$70 = -1.5(100) + b$$

$$b = 220$$

So, the equation of our line is:

$$y = -1.5x + 220$$

At what price would 140,000 lamps be bought by customers? (140, y)

$$y = -1.5(140) + 220$$

$$y = 10 \ (p \text{ in } \$)$$

Coordinate Geometry & Basic Shapes

Question 3: How many lamps would be bought by customers at $40/lamp price?

Solution:

We know that: $y = -1.5x + 220$

$$40 = -1.5x + 220$$

$$1.5x = 180$$

$$x = 120$$

Let's not forget that the number of lamps was in thousands;

So, $\qquad 120(1000) = 120,000 \; lamps$

Example 6.8

If $y = 3x$ and $y = \frac{x}{2}$ lines intersect the ABCD rectangle on points D and C, respectively, and A, B, C, D points are the corners of the ABCD rectangle, what is the perimeter of the ABCD rectangle given in the xy-plane below?

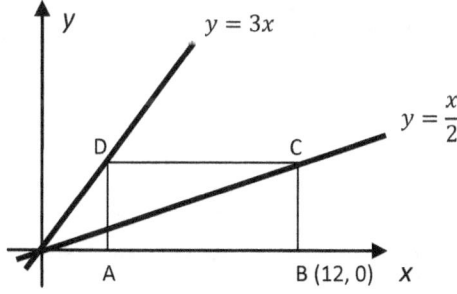

Solution:

  Point C has the same x value as B (12). We can also find the y value of point C.

$$y = \frac{x}{2}$$

$y = \frac{12}{2} = 6$ (for points B and C)

Since point C has the y value of 6, point D also has the same y value. We can now find the x value of point D since it is on $y = 3x$ line.

$6 = 3x$

$x = 2$ (for points A and D)

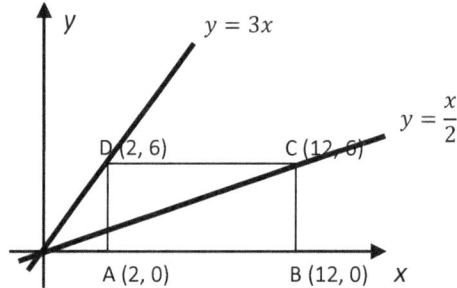

We see that:

$|AB| = |DC| = 12 - 2 = 10$

$|AD| = |BC| = 6 - 0 = 6$

Perimeter of ABCD Rectangle $= 2(10 + 6) = \mathbf{32}$

Example 6.9

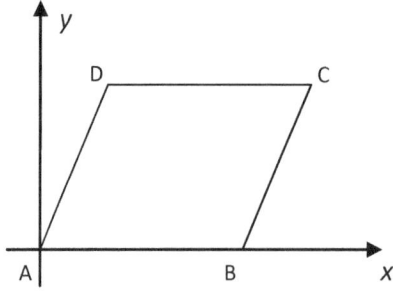

Coordinate Geometry & Basic Shapes

The ABCD Parallelogram above is lying on the x-axis and the coordinates of some of its corners are as follows:

$A = (0, 0), B = (5, 0), D = (3, 4)$

Find the perimeter of the ABCD Parallelogram and the sum of its diagonals.

Solution:

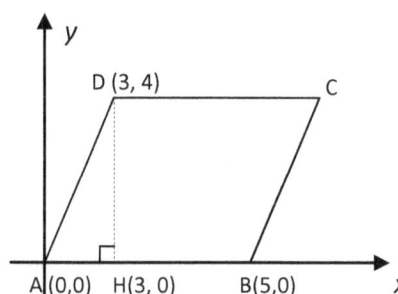

|AB| = 5

Since this is a parallelogram;

|AB| = |DC| = 5

We need to find |AD| or |CB|

Let's draw a perpendicular line from point D down to the |AB|, and define the touching point as H(3, 0).

Now, we have a right triangle (AHD Triangle) with sides: $|AH| = 3$, $|HD| = 4$ and $|AD| = ?$

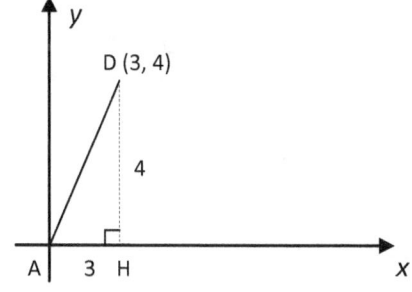

As per Pythagorean Theorem:

$|AH|^2 + |HD|^2 = |AD|^2$

$3^2 + 4^2 = |AD|^2$

$25 = |AD|^2$

$|AD| = 5$

So, the perimeter (P) of the ABCD Parallelogram is:

$P = |AB| + |DC| + |AD| + |BC|$

$P = 5 + 5 + 5 + 5$

$P = 20$

Now, let's find the length of its diagonals:

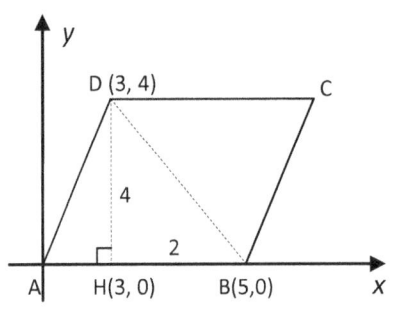

|DB| is one of the diagonals.

We know that $|AB| = 5$ and $|AH| = 3$

So; $|HB| = 2$

$|DB|^2 = 16 + 4$

$|DB| = 2\sqrt{5}$

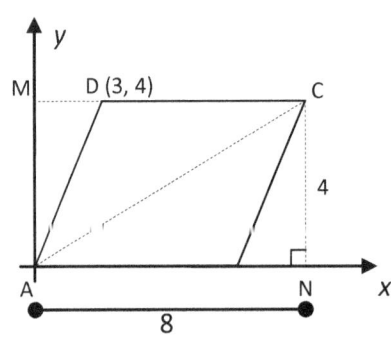

Let's have 2 more points: M and N.

Since $|MD| = 3$; and $|DC| = 5$;

$|MC| = 8 = |AN|$

Since $|DH| = 4$; $|CN| = 4$

$|AC|^2 = 64 + 16$

$|AC| = \sqrt{80}$

$|AC| = 4\sqrt{5}$

The sum of the ABCD Parallelogram's diagonals (Σ_d) is:

$\Sigma_d = |DB| + |AC|$

$\Sigma_d = 2\sqrt{5} + 4\sqrt{5}$

$\Sigma_d = 6\sqrt{5}$

Coordinate Geometry & Basic Shapes

Example 6.10

In the xy-plane below, if the lines *f* and *h* are perpendicular to each other, what are the coordinates of point B?

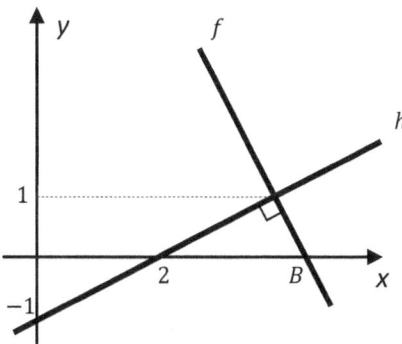

Solution:

Line *h* is passing through $(0, -1)$ and $(2, 0)$ points. Let's find its equation:

$m = \dfrac{y_2 - y_1}{x_2 - x_1}$

$m = \dfrac{0 - -1}{2 - 0} = \dfrac{1}{2}$

The slope of line *h*:

$(m_h) = \dfrac{1}{2}$

$y = mx + b$

$y = \dfrac{x}{2} + b$

Let's run the coordinates of one of our points, such as $(2, 0)$

$0 = \dfrac{2}{2} + b$

$b = -1$

So, the equation of line *h* is:

$$y = \frac{x}{2} - 1$$

Let's find the intersection point of lines *f* and *h*. (we know the *y* value which is 1)

$$y = \frac{x}{2} - 1$$

$$1 = \frac{x}{2} - 1$$

$$x = 4$$

> So the intersection point is (4, 1)

We know that lines *f* and *h* are perpendicular to each other.

So; $m_h \cdot m_f = -1$

$\frac{1}{2} \cdot m_f = -1$

$m_f = -2$

> When two lines are perpendicular to each other, the product of their slopes is always -1.

The intersection point (4, 1) is also on line *f*, and we know the slope of line *f* which is -2.

So, $m_f = \frac{y - y_1}{x - x_1}$

$-2 = \frac{y - 1}{x - 4}$

$y - 1 = -2x + 8$

$y = -2x + 9$

Finally, since point B is on *x-axis*, its *y* value is 0.

So; $0 = -2x + 9$

Coordinate Geometry & Basic Shapes

$$x = \frac{9}{2}$$

The coordinates of point B = $\left(\frac{9}{2}, 0\right)$

Example 6.11   Parabolas

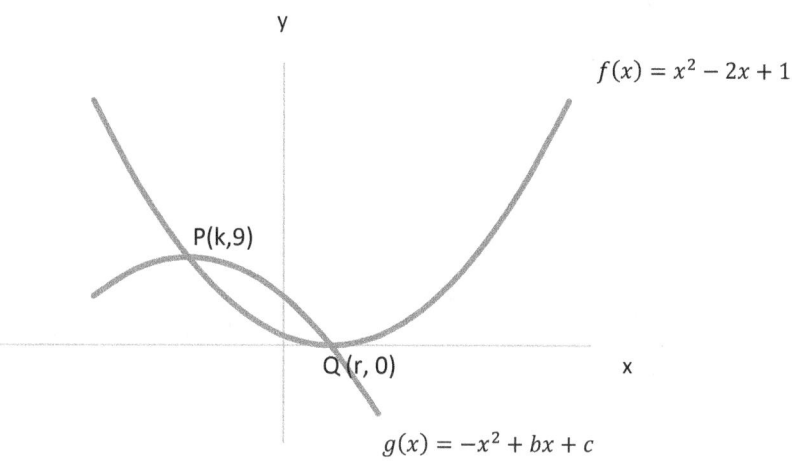

As shown in the xy-plane above, two parabolas intersect with each other at their vertices (P and Q). If one of the their intersection points is $P(k, 9)$, and the other one is $Q(r, 0)$; what is the value of $g(2)$?

Solution:

Let's start with the $f(x)$ parabola:

Reminder! The Vertex is the highest or lowest point of a parabola.

We need to find the vertex of the $f(x)$ parabola:

$$h = \frac{-b}{2a}$$

For $Q(h, 0)$ and $f(x) = ax^2 + bx + c$

$$f(x) = x^2 - 2x + 1$$

The vertex formula: $h = \frac{-b}{2a}$

$$h = \frac{2}{2} = 1$$

So; $Q(1, 0)$ is the vertex of $f(x)$. It also intersects with $g(x)$.

Since; $g(x) = -x^2 + bx + c$

$g(1) = -1^2 + b.1 + c$

$0 = -1 + b + c$

$b + c = 1$

We know that $P(k, 9)$ is on $f(x)$ parabola:

$f(x) = x^2 - 2x + 1$

$9 = k^2 - 2k + 1$

$k^2 - 2k - 8 = 0$

$(k + 2)(k - 4) = 0$

$k = -2 \; ; \; k = 4$

So; $P(-2, 9)$

> $P(k, 9)$ is located in the II. Quadrant of our xy-plane. Therefore: $k < 0$

We know that $P(-2, 9)$ is the vertex of $g(x)$ parabola.

$-2 = \dfrac{-b}{-2}$

$b = -4$

> For $P(-2,9)$ and $g(x) = -x^2 + bx + c$
>
> The vertex formula: $h = -2 = \dfrac{-b}{2a}$

We already found that: $b + c = 1$

$-4 + c = 1$

$c = 5$

So; $g(x) = -x^2 + bx + c$

$g(x) = -x^2 - 4x + 5$

$g(2) = -2^2 - 4.2 + 5$

$g(x) = -4 - 8 + 5$

$g(x) = -7$

Example 6.12

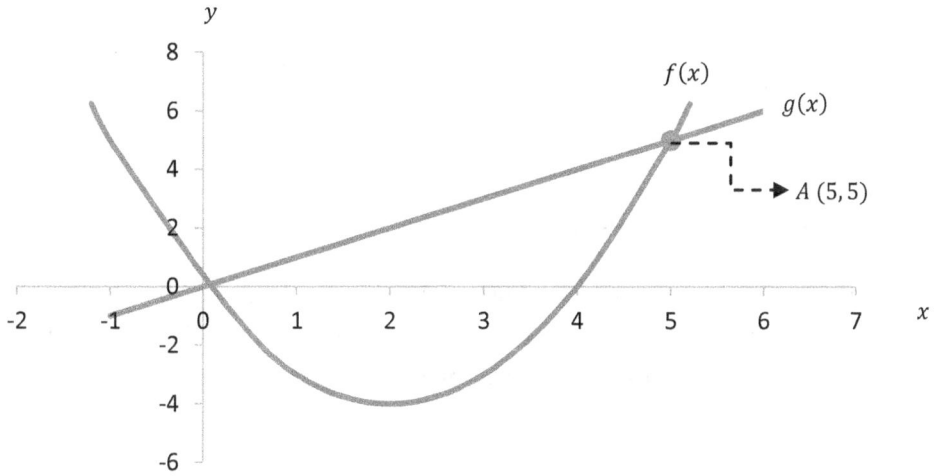

In the xy-plane above the axis of $f(x)$ parabola is parallel to the y-axis. If $A(5,5)$ and Origin $(0,0)$ are the intersection points of $f(x)$ parabola with $g(x)$ line, what is the value of $\dfrac{(f \circ g)(8)}{(f \circ f)(2)}$?

Solution:

We need to find the equations of both $f(x)$ parabola and $g(x)$ line.

We know the points of $g(x)$: $(5,5)$ and Origin $(0,0)$;

$g(x) = x$ and $g(8) = 8$

$$\dfrac{(f \circ g)(8)}{(f \circ f)(2)} = \dfrac{f(g(8))}{f(f(2))}$$

We know the points of $f(x)$: $(5, 5)$, Origin $(0, 0)$ and $(4, 0)$;

$$f(x) = x(x - 4) = x^2 - 4x$$

So, $f(8) = 8^2 - 32 = 32$

$f(2) = 2^2 - 8 = -4$

$f(-4) = (-4)^2 + 16 = 32$

$$\frac{(f \circ g)(8)}{(f \circ f)(2)} = \frac{f(g(8))}{f(f(2))} = \frac{f(8)}{f(-4)} = \frac{32}{32} = 1$$

Example 6.13 Shape Geometry

A logistics company receives an order of 180 identically same sized boxes to be delivered to another location. The length of the box is 4 feet, the width is 1 foot, and its height is 4 feet. The company uses same sized containers with dimensions as shown below:

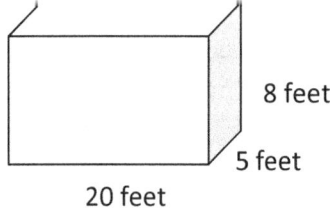

At least how many containers are needed to fit in 180 boxes?

Solution:

Let's find the Volume of each box: V_b

V . b = 4.1.4

V . b = 16 ft³

Now, let's find the Volume of each container: V_c

$V_c = 20.5.8$

$V_c = 800 \text{ ft}^3$

If we divide V_c by V_b, we will find how many boxes we can fit in one container.

800/16 = 50 (boxes can fit in one container)

We have a total of 180 boxes and we know that each container can accommodate up to 50 boxes.

$$180 = 50 + 50 + 50 + 30$$

So;   **minimum 4** containers are needed.

Practice Questions:

1. What is the number of degrees that the hour hand of a clock moves through between 1:00pm and 5:30pm of the same day?

2. In the xy-plane, the graph of a line passes through the point (2, 4) and crosses the x-axis at the point (4, 0). If the line crosses the y-axis at the point $(0, m)$, what is the value of m?

3. In the xy-plane, the graph of the equation below is a circle. Point A is on the circle and has coordinates $(-4, -8)$. If $\overline{AB}$ is a diameter of the circle, what are the coordinates of point B?
$$(x + 4)^2 + (y + 5)^2 = 9$$

4. The function g is defined by $g(x) = (x - 3)(x - 5)$. The graph of g in the xy-plane is a parabola. Which of the following intervals contains the x-coordinate of the vertex of the graph of g?
    a) $-2 < x < 2$
    b) $3 < x < 6$
    c) $0 < x < 4$
    d) $-3 < x < -5$

5. The graph below demonstrates the length (in feet) of distance, d, a product has travelled on a conveyor belt t minutes after the product was placed on the conveyor belt. If t and d relationship is always a line, and the product always gets placed at 1 feet length mark, at what length can the product be found 52 minutes after it was placed on the belt?

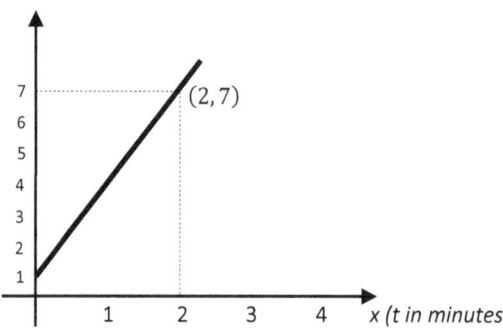

6.

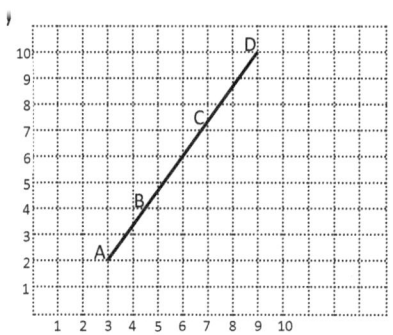

A, B, C and D are points on the same line shown in the xy-plane above. The coordinates of the points are: A (3, 2), B(..., 4), C(7, ...), D(9, 10)

What is the value of $\dfrac{|AC|}{|BD|}$?

7.

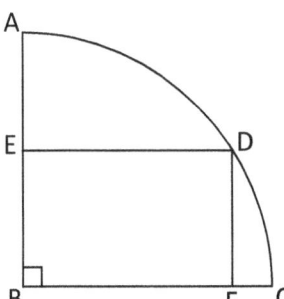

ABC is the quarter of a circle. EBFD is a rectangle with sides: $\overline{DF} = 12$ and $\overline{ED} = 16$. What is the length of the arc $\overset{\frown}{ADC}$?

8. What is the perimeter of the $DEFG$ rectangle drawn below?

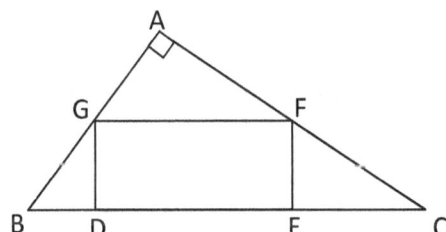

$\widehat{BAC} = 90°$

$|AG| = |BG|$

$|BD| = 1 \; ; \; |EC| = 4$

9. In the figure below; ABC and CDE are right triangles. If $|CB| = 4$, $|CD| = 3$, and $|AB| = 9$, what is the area of triangle ACE?

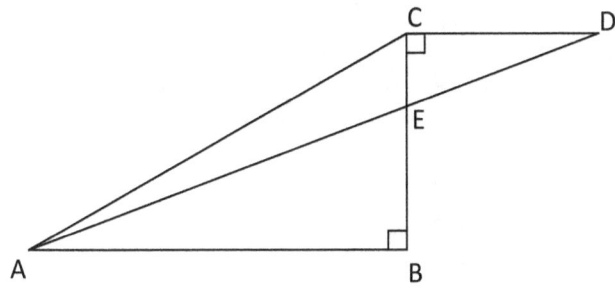

10. Which of the following could be the equation of the graph below?
    a) $y = (x - 3)(x + 4)(x - 1)$
    b) $y = x(x - 3)(x + 4)$
    c) $y = x^2(x - 3)(x + 4)$
    d) $y = x|(x^2 - 3)(x + 4)|$
    e) $y = x^3(x - 3)(x + 4)$

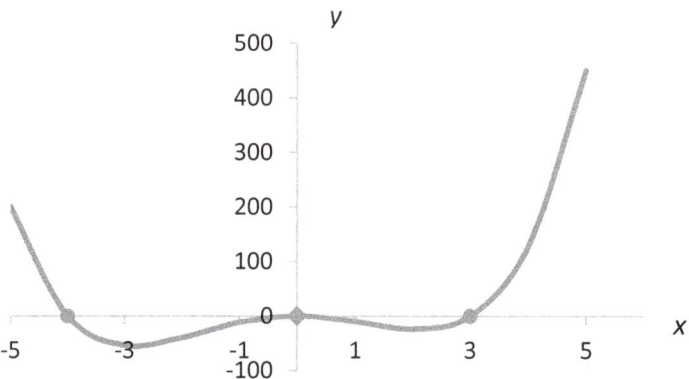

11. Which of the following could be the equation of the graph below?
    a) $y = (x - 3)(x + 2)(x - 1)$
    b) $y = x^2 - 4$
    c) $y = x^2 - 3$
    d) $y = x^3 - 1$
    e) $y = x^3 + 9$

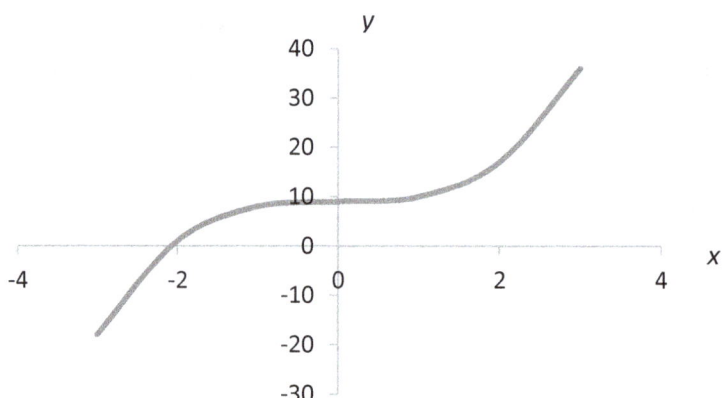

Coordinate Geometry & Basic Shapes

12. The *xy*-plane drawn below has been divided into 4 sections. If point $A(r, h)$ is located in section III, where is point $B(-h, r)$ located?

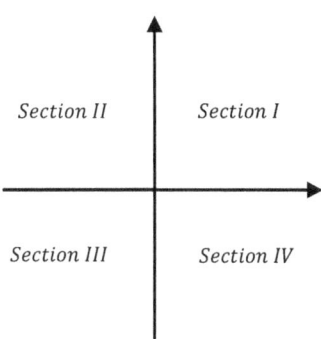

13. Which of the following could be the equation of the graph below?

   a) $y = |x - 1| - 1$
   b) $y = ||x| + 1| + 1$
   c) $y = ||x| - 1| + 1$
   d) $y = |x^2 - 1|$
   e) $y = |x - 1| + |x + 1|$

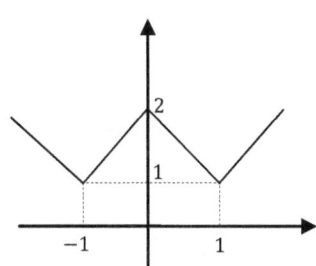

Key to Practice Questions

Chapter 1   Review of Algebra and Linear Equations

1) −1 2) 2 3) 4 4) 54 5) 9/5

6) 1 7) a 8) −7 9) 6 10) 6

11) 5 12) 11/3 13) −2 14) 5 15) 1

16) 2 17) 5 18) 4 19) 9 20) ¼

21) 5/4 22) 2/3 23) 6^{1-y} 24) $2\sqrt{2}$ 25) $\sqrt{3}$

26) 5 27) 10 28) 25/4 29) 1/2 30) 1/2

31) −12/5 32) $\frac{z+1}{z}$ 33) x 34) 4 35) {1, 5}

36) $\{\frac{5}{2}, \frac{-2}{3}\}$ 37) {−3, 5} 38) $\{\frac{\sqrt{5}}{2}, \sqrt{5}\}$ 39) {5, −2} 40) −5

41) 4 42) 2 43) $\frac{2}{a^2-1}$ 44) $\frac{3}{2(x+2k)}$ 45) x + y

46) $nx - 1$ 47) x 48) x + 1 49) $-2b$ 50) 3

51) −9 52) −5 53) 1 54) −2 55) −1

56) (2, −3) 57) −4 58) 835 59) 1/6 60) k/h

61) 8 62) −5 63) 28 64) 10 65) −3

66) 1/x 67) −8/27 68) 3 69) $5\left(\frac{1}{a^2} - \frac{1}{b^2}\right)$ 70) 1

71) 47 72) $5^y - 5^x$ 73) $64xy^2$ 74) 5 75) 22

76) 3 77) 4 78) 2 79) ¼ 80) 6

81) 8	82) 4	83) 9	84) −1	85) 12
86) 2	87) a/2	88) 4	89) 50	90) 2
91) 5	92) 10	93) 2	94) 1	

Chapter 2 — Problem Solving

1) 40 2) 60 3) 32% 4) 12,520 5) 10

6) $\frac{78x}{50}$ 7) C 8) 300 meters 9) 200 10) 8

11) 8:40am 12) 10 13) 16 14) $\$\frac{A+100}{50}$ 15) 0.675T

16) $15,240 17) 10% 18) $35,669 19) $812.5 20) $1322

21) 16 22) D=2B/3 23) $72,000 24) 0.8 lt. 25) 10%

26) 100% 27) 200 28) $30 29) 55 30) 4%

31) 2% loss 32) 4% decline 33) 80 34) 9% decline 35) 8

36) 20 37) $4x = y + z$ 38) 13.3% 39) 35

Chapter 3 — Chart, Table and Data Analysis

1) C 2) C & D 3) C 4) B 5) B

6) D 7) 291,500 8) 592,280 9) 25% 10) 6.98%

11) Apr-May 12) B & {5, 2.01} 13) 5 14) 175cm 15) B

16) 8 17) 4

Key to Practice Questions

Chapter 4 — Functions

1) 4
2) 10
3) 3
4) −2
5) 1
6) −47/15
7) 43,700
8) $LV(n) = X + n(Y - Z)$
9) 4.3
10) 26
11) 1
12) 14/5
13) $\frac{x^2}{9}$
14) 15
15) D
16) 2/3
17) 9
18) −1.75 or −7/4

Chapter 5 — Inequalities, Statistics and Complex Numbers

1) −8
2) $-1 < x < 3$
3) $1/3 < x < 5$
4) 0
5) $x \leq 0; x > 3$
6) $x < -3; 0 < x < 2$
7) −4
8) 1
9) 20
10) 14
11) −2
12) $-4i$
13) $7.2 million
14) 26
15) B

Chapter 6 — Coordinate Geometry and Basic Shapes

1) 135°
2) 8
3) $(-4, -2)$
4) B
5) 157 feet
6) 8/9
7) 10π
8) 14
9) 4.5
10) C
11) E
12) Sec IV
13) C

www.ingramcontent.com/pod-product-compliance
Lightning Source LLC
Chambersburg PA
CBHW082334220526
45470CB00008B/2502